Evaluation of Guidelines for Exposures to Technologically Enhanced Naturally Occurring Radioactive Materials

Committee on Evaluation of EPA Guidelines for Exposure to
Naturally Occurring Radioactive Materials
Board on Radiation Effects Research
Commission on Life Sciences
National Research Council

NATIONAL ACADEMY PRESS
Washington, DC 1999

NATIONAL ACADEMY PRESS • 2101 CONSTITUTION AVENUE, N.W. • WASHINGTON, D.C. 20418

NOTICE: The project that is the subject of this report was approved by the Governing Board of the National Research Council, whose members are drawn from the councils of the National Academy of Sciences, the National Academy of Engineering, and the Institute of Medicine. The members of the committee responsible for the report were chosen for their special competences and with regard to appropriate balance.

This report was prepared under Environmental Protection Agency contract 68D70009 between the National Academy of Sciences and the Environmental Protection Agency.

Library of Congress Number 98-83159
International Standard Book Number 0-309-06297-7

A limited number of copies of this report are available from

National Research Council
Board on Radiation Effects Research
Room 342
2101 Constitution Avenue, NW
Washington, DC 20418
(202) 334-2232

Copyright © 1999 by the National Academy of Sciences. All rights reserved.

Printed in the United States of America

COMMITTEE ON EVALUATION OF EPA GUIDELINES FOR EXPOSURE TO NATURALLY OCCURRING RADIOACTIVE MATERIALS

BERNARD D. GOLDSTEIN *(Chair)*, UMDNJ-Robert Wood Johnson Medical School, Piscataway, NJ
MERRIL EISENBUD, Chapel Hill, NC (deceased August 1997)
THOMAS F. GESELL, Idaho State University, Pocatello, ID
SHAWKI A. IBRAHIM, Colorado State University, Fort Collins, CO
DAVID C. KOCHER, Oak Ridge National Laboratory, Oak Ridge, TN
EDWARD R. LANDA, US Geological Survey, Reston, VA
ANSELMO S. PASCHOA, Pontifical Catholic University of Rio de Janeiro, Rio de Janeiro, Brazil

CLS ADVISER

FREDERICK R. ANDERSON, Cadwalader, Wickersham and Taft, Washington, D C

NATIONAL RESEARCH COUNCIL STAFF

STEVEN L. SIMON, Study Director, Board on Radiation Effects Research
KAREN M. BRYANT, Project Assistant (until 10/23/98)
DORIS E. TAYLOR, Staff Assistant
NORMAN GROSSBLATT, Editor

SPONSOR'S PROJECT OFFICER

LOREN W. SETLOW, US Environmental Protection Agency

BOARD ON RADIATION EFFECTS RESEARCH

JOHN B. LITTLE *(Chair)*, Harvard School of Public Health, Boston, MA (until 6/30/98)
R.J. MICHAEL FRY, Oak Ridge, TN*
S. JAMES ADELSTEIN, Harvard Medical School, Boston, MA**
VALERIE BERAL, University of Oxford, United Kingdom
EDWARD R. EPP, Harvard University, Boston, MA**
HELEN H. EVANS, Case Western Reserve University, Cleveland, OH**
MERRIL EISENBUD, Chapel Hill, NC (deceased August 1997)
MAURICE S. FOX, Massachusetts Institute of Technology, Cambridge, MA
PHILIP C. HANAWALT, Stanford University, Palo Alto, CA (member until 6/30/98)
LYNN W. JELINSKI, Cornell University, Ithaca, NY
WILLIAM F. MORGAN, University of California, San Francisco**
WILLIAM J. SCHULL, The University of Texas Health Science Center, Houston, TX
DANIEL O. STRAM, University of Southern California, Los Angeles, CA
SUSAN W. WALLACE, University of Vermont, Burlington, Vermont (until 6/30/98)
H. RODNEY WITHERS, UCLA Medical Center, Los Angeles, California

NATIONAL RESEARCH COUNCIL STAFF

EVAN B. DOUPLE, Director, Board on Radiation Effects Research
RICK JOSTES, Senior Program Officer
STEVEN L. SIMON, Senior Program Officer
CATHERINE S. BERKLEY, Administrative Associate
KAREN BRYANT, Project Assistant (until 10/23/98)
PEGGY JOHNSON, Project Assistant (until 8/19/98)
DORIS E. TAYLOR, Staff Assistant

* New BRER Chair effective 7/1/98
** New members effective 7/1/98

COMMISSION ON LIFE SCIENCES

THOMAS D. POLLARD *(Chair)*, The Salk Institute for Biological Studies, La Jolla, CA
FREDERICK R. ANDERSON, Cadwalader, Wickersham & Taft, Washington, DC
JOHN C. BAILAR, III, University of Chicago, IL
PAUL BERG, Stanford University School of Medicine, Palo Alto, CA
JOANNA BURGER, Rutgers University, Piscataway, NJ
SHARON L. DUNWOODY, University of Wisconsin, Madison, WI
JOHN L. EMMERSON, Indianapolis, IN
NEAL L. FIRST, University of Wisconsin, Madison, WI
URSULA W. GOODENOUGH, Washington University, St. Louis, MO
HENRY W. HEIKKINEN, University of Northern Colorado, Greeley, CO
HANS J. KENDE, Michigan State University, East Lansing, MI
CYNTHIA J. KENYON, University of California, San Francisco, CA
DAVID M. LIVINGSTON, Dana-Farber Cancer Institute, Boston, MA
THOMAS E. LOVEJOY, Smithsonian Institution, Washington, DC
DONALD R. MATTISON, University of Pittsburgh, Pittsburgh, PA
JOSEPH E. MURRAY, Wellesley Hills, MA
EDWARD E. PENHOET, Chiron Corporation, Emeryville, CA
MALCOLM C. PIKE, Norris/USC Comprehensive Cancer Center, Los Angeles, CA
JONATHAN M. SAMET, The Johns Hopkins University, Baltimore, MD
CHARLES F. STEVENS, The Salk Institute for Biological Studies, La Jolla, CA
JOHN L. VANDEBERG, Southwest Foundation for Biomedical Research, San Antonio, TX

NATIONAL RESEARCH COUNCIL STAFF

PAUL GILMAN, Executive Director
ALVIN G. LAZEN, Associate Executive Director

The National Academy of Sciences is a private, nonprofit, self-perpetuating society of distinguished scholars engaged in scientific and engineering research, dedicated to the furtherance of science and technology and to their use for the general welfare. Upon the authority of the charter granted to it by the Congress in 1863, the Academy has a mandate that requires it to advise the federal government on scientific and technical matters. Dr. Bruce M. Alberts is president of the National Academy of Sciences.

The National Academy of Engineering was established in 1964, under the charter of the National Academy of Sciences, as a parallel organization of outstanding engineers. It is autonomous in its administration and in the selection of its members, sharing with the National Academy of Sciences the responsibility for advising the federal government. The National Academy of Engineering also sponsors engineering programs aimed at meeting national needs, encourages education and research, and recognizes the superior achievements of engineers. Dr. William A. Wulf is the president of the National Academy of Engineering.

The Institute of Medicine was established in 1970 by the National Academy of Sciences to secure the services of eminent members of appropriate professions in the examination of policy matters pertaining to the health of the public. The Institute acts under the responsibility given to the National Academy of Sciences by its congressional charter to be an adviser to the federal government and, upon its own initiative, to identify issues of medical care, research, and education. Dr. Kenneth I. Shine is president of the Institute of Medicine.

The National Research Council was organized by the National Academy of Sciences in 1916 to associate the broad community of science and technology with the Academy's purposes of furthering knowledge and advising the federal government. Functioning in accordance with general policies determined by the Academy, the Council has become the principal operating agency of both the National Academy of Sciences and the National Academy of Engineering in providing services to the government, the public, and the scientific and engineering communities. The Council is administered jointly by both Academies and the Institute of Medicine. Dr. Bruce M. Alberts and Dr. William A. Wulf are chairman and vice chairman, respectively, of the National Research Council.

PREFACE

The human propensity to alter our environment has frequently led to the shifting of the earth's crustal constituents, at times moving naturally occurring radioactive materials (NORM) in closer proximity to ourselves, and at other times increasing human radiation exposure by enriching the concentration of technologically enhanced naturally occurring radioactive materials (TENORM).

As a result of a request from the 104th Congress (Section 311 of H.R. 2099 and Senate Committee Report 104-318 of H.R. 3666), the US Environmental Protection Agency (EPA) asked the National Research Council to investigate and report on the scientific bases for the public recommendations of the EPA with respect to indoor radon and other naturally occurring radioactive materials. Specifically, the Research Council was asked to address the question of whether the differences in guidelines related to NORM and developed by EPA and other organizations are based upon scientific and technical information or on risk management policy. The Reseach Council was asked to comment on the relative merit of any scientific or technical differences and to assess whether there is relevant scientific information that has not been used in the development of the guidelines for NORM.

The study began in March of 1997 and a committee of six scientists including international representation was appointed by the National Research Council to provide the answers to the specific EPA requests. The committee met 5 times to gather information and to deliberate its findings. Deeply saddened by the death of one of its members, the committee dedicates its report to that important member, Dr. Merril Eisenbud.

During the course of the committee's deliberations, several individuals provided information to the committee. Appreciation for these contributions is extended to the following:

Joseph Alvarez, Auxier and Associates
Jean-Claude Dehmel, ANSI/HPS NORM Standard Working Group
William P. Dornsife, Conference of Radiation Control Program Directors
Naomi Harley, New York University Medical Center
Joseph Hezir, EOP Group

Corey McDaniel, EOP Group
Christopher B. Nelson, Environmental Protection Agency
William A. Mills, Olney, MD
Robert A. Nelson, Nuclear Regulatory Commission
Marty Reape, FMC Corporation
Alan B. Richardson, Environmental Protection Agency
Loren W. Setlow, Environmental Protection Agency
Charles Simmons, Kilpatrick & Stockton, LLP
Robert Simon, Senate Committee on Energy and Natural Resources
Phyllis Sobel, Nuclear Regulatory Commission
Lawrence G. Weinstock, Environmental Protection Agency

This report has been reviewed in draft form by individuals chosen for their diverse perspectives and technical expertise, in accordance with procedures approved by the National Research Council's Report Review Committee. The purpose of this independent review is to provide candid and critical comments that will assist the institution in making the published report as sound as possible and to ensure that the report meets institutional standards for objectivity, evidence, and responsiveness to the study charge. The review comments and draft manuscript remain confidential to protect the integrity of the deliberative process. We wish to thank the following individuals for their participation in the review of this report:

Frederick R. Anderson, Cadwalader, Wickersham & Taft
Raymond D. Cooper, St. Petersburg, FL
Richard J. Guimond, Motorola, Inc.
William A. Mills, Olney, MD
Dade W. Moeller, Dade Moeller & Associates, Inc.
Raymond Paris, Oregon Health Division
Richard B. Setlow, Brookhaven National Laboratory
Charles Simmons, Kilpatrick & Stockton, LLP

While the individuals listed above have provided constructive comments and suggestions, it must be emphasized that responsibility for the final content of this report rests entirely with the authoring committee and the institution.
 The committee is very appreciative for the expertise, dedication, and hard work of the study director, Dr. Steven L. Simon. The attention to administrative details by Karen Bryant and Doris Taylor from the Research Council's Board on Radiation Effects Research, is also appreciated.

Bernard D. Goldstein, *Chairman*

DEDICATION
In memory of *Merril Eisenbud* (1915 – 1997)

This report is in many ways a product of the career of Professor Merril Eisenbud. Not only was Merril Eisenbud a member of the committee until his death in August 1997, but his scholarly writings in the discipline of environmental radiation were important influences for several generations of students and collaborators, some of whom have served on this committee. Merril Eisenbud, a truly great scientist and public-health visionary, was one of the first to actively study and teach the environmental-health implications of naturally occurring and human-made radioactive substances.

Merril Eisenbud's university studies and remarkable scientific career spanned the period from the discovery of the neutron in 1932, through the development of nuclear technology, and finally to the cleanup of the world's nuclear-weapons complexes. He contributed actively long after the normal age of retirement; he published a book and several journal articles in his last year.

Merril began his working career as an industrial hygienist for the Liberty Mutual Insurance Company, engaging in studies of chemical and radiation hazards in industry from 1936 to 1947. His remarkable talents and energy allowed him to contribute prolifically as a scientist throughout his career while holding several demanding managerial positions. These jobs included 12 years (1947-1959) with the US Atomic Energy Commission, where he was the founding director of the Health and Safety Laboratory[1]. From 1954 to 1959, he served in a dual capacity as laboratory director and manager of the AEC New York Operations Office. For 2 years (1968-1970), he served as the first environmental-protection administrator for the City of New York.

Merril Eisenbud's university teaching career began in 1959 when he joined the New York University Medical Center's Institute of Environmental Medicine as professor and director of the Laboratory of Environmental Studies. On retirement from active teaching at NYU in 1984 he continued on as professor emeritus of environmental medicine. At the time of his death, he was also distinguished scholar in residence at the Duke University Medical Center and adjunct professor of environmental sciences and engineering at the University of North Carolina School of Public Health.

---

[1]Now the Environmental Measurements Laboratory of the US Department of Energy

Merril held a BSEE from the New York University College of Engineering and two honorary doctoral degrees in science. He was a member of many national and international committees, including those of agencies of the United Nations, the National Research Council, and the US government. He had been a member of the advisory councils of the Electric Power Research Institute, the Institute of Nuclear Power Operations, and the Beryllium Industry Scientific Advisory Committee. He was serving the National Research Council as a member of its Board on Radiation Effects Research at the time of his death.

Among the awards received by Merril were the Hermann M. Biggs Medal of the New York State Public Health Association, the Arthur H. Compton Award of the American Nuclear Society, the Gold Medal of the US Atomic Energy Commission, the Distinguished Achievement Award of the Health Physics Society, the Life Award of the Power Division of the Institute of Electronic and Electrical Engineers, and the Taylor Medal of the National Council on Radiation Protection and Measurements. He was an honorary life fellow of the New York Academy of Sciences, a member of the National Academy of Engineering, a corresponding member of the Brazilian Academy of Sciences, and a fellow of the New York Academy of Medicine.

Merril published prolifically on environmental radioactivity, urban pollution, environmental effects of power generation, and human ecology. His books include four editions of *Environmental Radioactivity*, the most recent published in 1997; his autobiography, *An Environmental Odyssey* (1990); *The Environment, Technology, and Health: Human Ecology in Historic Perspective* (1978); and *Biological Effects of Electric and Magnetic Fields of Extremely Low Frequency* (1977). He contributed more than 200 journal articles and book chapters to the scientific literature.

This report is dedicated to Merril Eisenbud, our friend, mentor, and colleague.

Contents

Executive Summary		1
1.	Introduction	16
2.	Natural Radioactivity and Radiation	25
3.	Major Sources of Technologically Enhanced Naturally-Occurring Radioactive Materials	62
4.	Role of Exposure and Dose or Risk Assessments in Developing Radiation Standards	77
5.	Basic Approaches to Regulating Radiation Exposures of the public	92
6.	Organizations Concerned With Radiation Protection of the Public	100
7.	Environmental Protection Agency Guidances and Regulations for Naturally Occurring Radionuclides	109
8.	Indoor-Radon Guidelines and Recommendations	161
9.	Other Guidances for TENORM	188
10.	Comparison of Current Guidances for TENORM in the Environment	209
11.	Issues in Developing Guidances for TENORM	224
12.	Conclusions and Recommendations	250

CONTENTS (CONTINUED)

REFERENCES

APPENDIX
    Radiation Quantities and Units
    Definitions
    Acronyms

INFORMATION ON COMMITTEE MEMBERS

Executive Summary

INTRODUCTION

Naturally occurring radionuclides are found throughout the earth's crust, and they form part of the natural background of radiation to which all humans are exposed. Many human activities—such as mining and milling of ores, extraction of petroleum products, use of groundwater for domestic purposes, and living in houses—alter the natural background of radiation either by moving naturally occurring radionuclides from inaccessible locations to locations where humans are present or by concentrating the radionuclides in the exposure environment. Such alterations of the natural environment can increase, sometimes substantially, radiation exposures of the public.

Exposures of the public to naturally occurring radioactive materials (NORM) that result from human activities that alter the natural environment can be subjected to regulatory control, at least to some degree. The regulation of public exposures to such technologically enhanced naturally occurring radioactive materials (TENORM) by the US Environmental Protection Agency (EPA) and other regulatory and advisory organizations is the subject of this study by the National Research Council's Committee on the Evaluation of EPA Guidelines for Exposures to Naturally Occurring Radioactive Materials.

The committee has used the term *technologically enhanced naturally occurring radioactive materials* to refer to the materials of concern to this study and has defined this term as follows:

> Technologically enhanced naturally occurring radioactive materials are any naturally occurring radioactive materials not subject to regulation under the Atomic Energy Act whose radionuclide concentrations or potential for human exposure have

been increased above levels encountered in the natural state by human activities.

The exclusion of NORM subject to Atomic Energy Act jurisdiction from the definition of TENORM means that this study is not concerned with evaluating guidelines developed by EPA or other federal agencies that apply to NORM associated with the production and use of nuclear fuels, including uranium and thorium mill tailings, naturally occurring radionuclides released to the environment during operations of nuclear fuel-cycle facilities, or natural uranium or thorium in the form of source material. The most important radionuclides in TENORM as defined in this study include the long-lived, naturally occurring isotopes of radium, thorium, and uranium and their radiologically important decay products (such as radon), as well as potassium-40.

PURPOSE AND SCOPE OF STUDY

This study resulted from considerations by EPA and other organizations of guidelines for controlling exposures of the public to TENORM. Specifically, there has been a concern that EPA and other organizations have arrived at different numerical values for guidelines for essentially the same exposure situations but that the reasons for the differences, especially the extent to which they were based on scientific and technical considerations, were not apparent.

In light of that concern, the National Research Council committee was asked to address the following questions:

- Whether the differences in the guidelines for TENORM developed by EPA and other organizations are based upon scientific and technical information, or on policy decisions related to risk management.

- If the guidelines developed by EPA and other organizations differ in their scientific and technical bases, what the relative merits of the different scientific and technical assumptions are.

- Whether there is relevant and appropriate scientific information that has not been used in the development of contemporary risk analysis for NORM.

EXECUTIVE SUMMARY

In view of the concern about guidelines for TENORM and their scientific and technical bases, the committee has considered approaches to risk assessment for NORM only in regard to their use in developing and implementing guidelines for controlling radiation exposures of the public. The committee has not addressed other issues of risk assessment for NORM that may arise in attempting to estimate real risks posed by actual exposures of individuals or populations.

The guidelines evaluated in this study include those for indoor radon and those for any other TENORM. Those two types of NORM generally have been regulated separately. After a thorough review of the subject, we have concluded that the differences between regulatory agencies' and other organizations' guidances for the control of TENORM have little basis in science but reflect differences in risk-management approaches and organizational missions.

This study is not concerned with evaluations of nonscientific issues of importance to the development of guidelines for NORM, such as costs and policy judgments in risk management. However, in evaluating whether differences in guidelines for NORM developed by EPA and other organizations are based on scientific and technical information, the committee found it necessary to identify important policy judgments in risk management that have influenced development of the guidelines, even though the merit of any such judgments is not considered.

The EPA's current guidelines for indoor radon, however, are easily identified. In the case of TENORM other than indoor radon, it was not a simple matter for the committee to define what the current EPA guidelines are. Some existing guidelines clearly are outdated and do not represent EPA's current views on suitable approaches to regulating TENORM; proposals for revising some of them have been published, but the proposals have not been issued in final form, so there is uncertainty about what the new guidelines might be. Some guidelines are in the form of legally enforceable regulations, but EPA's preferred approach to regulating TENORM in some cases is indicated only by policy statements, and there are no published guidelines for some important exposure situations. Thus, judgment was required by the committee in selecting EPA guidelines for TENORM other than indoor radon to be emphasized in the comparisons with similar guidelines developed by other organizations.

In general, the committee has emphasized the most recent statements by EPA concerning guidelines for TENORM other than indoor radon, regardless of their form or status. However, the committee also has attempted to discuss all EPA guidelines, in whatever form, in an effort to provide a reasonably complete picture of EPA's current policies.

RESPONSES TO COMMITTEE CHARGE

The following sections summarize the committee's responses to the three parts of its charge.

Technical Basis for Differences in Guidelines for TENORM

The committee has reviewed existing or proposed guidelines for TENORM developed by EPA and similar guidelines developed by other regulatory or advisory organizations in the United States and elsewhere. The other organizations whose guidelines have been considered are the National Council on Radiation Protection and Measurements (NCRP), the International Commission on Radiological Protection (ICRP), other federal agencies, state agencies and organizations, regulatory authorities in other nations, and other national and international advisory organizations. In keeping with its charge, the committee paid particular attention to the bases of the various guidelines developed by the different organizations.

There clearly are differences in the numerical values of the most recent guidelines for TENORM developed by EPA and some of the guidelines for similar exposure situations developed by other organizations. Differences are found in the guidelines for indoor radon and for TENORM other than indoor radon. Furthermore, where there are differences, EPA guidelines tend to be more restrictive, that is to correspond to lower levels of exposure and therefore presumably lower risks to the public. Whether the differences between the EPA guidelines and those developed by other organizations are significant is entirely a matter of judgment.

On the basis of its review, <u>the committee finds that the differences between EPA guidelines for TENORM and similar guidelines developed by other organizations are not based on scientific and technical information.</u> This conclusion is based primarily on the following considerations:

- All organizations that have developed guidelines for indoor radon have assumed approximately the same risk associated with exposure to radon and its short-lived decay products based on epidemiologic data obtained from studies of underground miners and extrapolation of these data to exposures to radon in indoor residences.

- All organizations that have developed guidelines for TENORM other than indoor radon have assumed approximately the same risk associated with uniform irradiation of the whole body based on epidemiologic data obtained primarily from studies of the Japanese atomic-bomb survivors and extrapolation of these data to the low doses of concern in environmental exposures.

- All organizations that have developed guidelines for TENORM have assumed a linear, no-threshold dose-response relationship at low levels of exposure.

Thus, the committee finds that differences in the guidelines for TENORM developed by EPA and other organizations are based essentially on differences in policy judgments for risk management. That is not to say that EPA and other organizations have used the same methods and assumptions in estimating risks posed by radiation exposure. For indoor radon, different organizations have assumed somewhat different lifetime risks of lung cancer associated with exposure to short-lived radon decay products in air based, for example, on differences in the assumed risk-projection models, and the risk estimates have changed over time. EPA also has given greater attention than other organizations to the dependence of lung-cancer risk on an individual's smoking history.

Similarly, for radionuclides other than radon, EPA has used methods and assumptions that differ from those normally used by other organizations in estimating risks associated with chronic lifetime exposure. As a result, EPA's current estimate of the risk posed by external exposure is slightly higher than the risk estimate currently used by most other organizations, but EPA's risk estimates for internal exposure to the important long-lived alpha-emitting radionuclides in TENORM are, in some cases, substantially lower than risk estimates obtained with the methods and assumptions of other organizations. An example is the risk from ingestion or inhalation of thorium.

However, the differences between EPA guidelines for TENORM and the guidelines developed by other organizations are not a reflection of differences in the methods and assumptions for risk assessments for radon and other naturally occurring radionuclides. That is, EPA's current approach to risk assessment, as it differs from the approaches normally used by other organizations, was not an important factor in developing the numerical values of its current guidelines for TENORM.

Relative Merits of Different Scientific and Technical Assumptions

Given that the differences between the guidelines for TENORM developed by EPA and other organizations are not, in the committee's opinion, based on scientific information, the second question in the charge to the committee is moot. However, the committee has considered the differences between EPA's current methods and assumptions for risk assessment and the approaches normally used by other organizations, even though these differences have not had a substantial influence on the development of guidelines for TENORM. The committee's views on the approaches to risk assessment are summarized later.

Development of Contemporary Risk Analysis for NORM

The third part of the charge to the committee was to consider whether there is relevant and appropriate scientific information that has not been used in the development of contemporary risk analysis for NORM, especially risk analysis for purposes of developing and implementing guidelines for radiation exposure. A particular concern expressed to the committee is that some of the important radionuclides are parents of long decay chains involving a complex mixture of radioisotopes of different chemical elements and that exposures to such mixtures might necessitate novel approaches to methods of risk estimation.

The committee is not aware of any evidence that the properties of NORM differ from the properties of any other radionuclides in ways that would necessitate the development of different approaches to risk assessment. In regard to radiological properties, if one accepts the view currently held by all regulatory and advisory organizations involved in radiation protection that estimates of absorbed dose in tissue are the fundamental physical quantities that determine radiation risks for any exposure situation, there is no plausible rationale for any differences in risks due to ionizing radiation arising from naturally occurring and any other radionuclides, because absorbed dose in tissue depends only on the radiation type and its energy, not on the source of the radiation.

The decay chains of some naturally occurring radionuclides are considerably more complex than the decay chains of other radionuclides with regard to the number of decay products and chemical elements involved. However, contemporary methods of risk assessment that estimate doses and risks related to ingestion or inhalation of radionuclides by assuming that decay products produced in the body are redistributed and retained in the body according to the metabolic behavior characteristic of particular chemical elements take the added complexity into account by using the same methods that are applied to other radionuclides with many fewer decay products.

Thus, in general, there should be no difference between NORM and other radioactive materials with regard to suitable approaches to estimating doses and risks related to external or internal exposure. However, because naturally occurring radionuclides are ubiquitous in the exposure environment, there might be more opportunity than there is with many human-made radionuclides to use observational data on natural levels in different environmental compartments (such as soil, water, air, plants, and animals) and the fluxes between compartments to calibrate exposure-pathway models for TENORM. On the other hand, the ability to use such natural analog data for exposure pathway analysis must be tempered by the recognition that the physical and chemical forms of TENORM could be substantially different from those for the same elements in the natural environment. In that case, observations on the behavior of radionuclides in natural systems might not be relevant to the exposure situation of concern.

OTHER CONCLUSIONS AND RECOMMENDATIONS

During this study, the committee considered other issues related to the development of guidelines for TENORM. The committee's views on some of these issues are summarized below.

Policy Judgments for Risk Management

The committee has concluded that the differences between EPA guidelines for TENORM and similar guidelines developed by other organizations are based essentially on differences in policy judgments for risk management, rather than differences in scientific and technical information. An evaluation of the relative merit of the differences in policy judgments for risk management was not part of this study, but the committee needed to identify these judgments in reaching the conclusion that the differences in the guidelines do not reflect differences in scientific and technical information.

From its considerations of the various guidelines for TENORM developed by EPA and other organizations, the committee believes that the differences often are based, at least in part, on two factors that are strictly matters of policy:

- Differences in judgments about acceptable risks related to exposure to TENORM.

- Differences in judgments about levels of TENORM in the environment, doses, or risks that are reasonably achievable.

Judgments about what is reasonably achievable in controlling exposures to TENORM take into account such factors as the costs of reducing exposures in relation to the benefits in reduced health risks to the public and other societal concerns. A particularly important consideration in regulating TENORM is the pre-existing levels of naturally occurring radionuclides in the environment and associated human exposures.

An additional factor that has been important in developing guidelines for TENORM is a judgment about the extent to which existing guidelines for particular exposure situations can be transferred to other situations. For example, some organizations have developed guidelines for TENORM in soil based on the concentration limits in current EPA guidelines for cleanup of soil contaminated with radium at uranium mill tailings sites developed under the Atomic Energy Act. <u>Transferability of standards developed for a specific class of TENORM waste is limited by the extent that the physical and chemical properties of the TENORM in issue, as well as projected exposure pathways, are substantially similar to those considered for uranium mill tailings.</u>

Other policy judgments for risk management also have been important causes of the differences between EPA guidelines and guidelines developed by other organizations. Some guidelines are concerned primarily with reducing risks to individuals who receive the highest exposures, and others with reducing risks to whole populations. Some guidelines include exposures to natural background, and others do not. Finally, in accordance with legislative mandates, some EPA guidelines apply only to a particular environmental medium (such as air) or a particular exposure pathway (such as drinking water), whereas guidelines developed by most other organizations, especially those expressed in terms of dose or risk, apply to all environmental media and all exposure pathways combined.

The importance of differences in policy judgments for risk management in determining differences in guidelines for TENORM is illustrated by the following three examples.

First, the EPA guideline for mitigation of radon in homes, 150 Bq/m^3 (4 pCi/L), is lower than mitigation levels recommended by NCRP and ICRP. The differences result from differences in the primary focus of the guidelines. NCRP and ICRP were concerned primarily with mitigation of risks to individuals who receive the highest exposures, and their recommended action levels were based primarily on judgments about the maximum tolerable risk, with some consideration of the feasibility of achieving concentrations below the action levels. EPA also was concerned with limiting individual risks but emphasized reducing exposures in the greatest number of homes possible, on the basis of judgments about levels that were reasonably achievable in most homes.

Second, EPA has issued proposed guidance on radiation protection of the public that includes a limit on annual dose equivalent of 1 mSv from all controlled sources combined, including TENORM, naturally occurring radionuclides from the nuclear fuel cycle, and human-made radionuclides but excluding radon. In contrast, NCRP has recommended an annual dose equivalent of 5 mSv as a remedial-action level for all natural sources, including natural background and TENORM but excluding radon. EPA's annual dose limit of 1 mSv for all controlled sources combined normally should be considerably more restrictive than NCRP's recommended remedial-action level of 5 mSv for all natural sources.

The difference between the two guidelines is due entirely to differences in policy judgments for risk management because EPA and NCRP assumed the same risk per unit dose. The most important difference is in the judgments about the maximum tolerable risk posed by exposure to TENORM. EPA regards TENORM other than indoor radon as a type of controlled source that should be regulated in the same manner as human-made sources, whereas NCRP essentially regards sources of TENORM, even if they are controllable, as a form of natural background that should be controlled differently from human-made sources. The other important difference in policy judgments is that EPA's dose limit excludes the dose from undisturbed natural background, whereas NCRP's remedial-action level includes the dose from background.

The third example concerns guidelines for cleanup of radioactively contaminated sites. The Nuclear Regulatory Commission has issued regulations that specify that sites are acceptable for unrestricted use if the annual dose equivalent from all exposure pathways, including the use of groundwater as a source of drinking water, does not exceed 0.25 mSv. EPA has objected to the Nuclear Regulatory Commission standards on two grounds. First, EPA believes that the annual dose from all exposure pathways should be limited to 0.15 mSv to achieve an acceptable level of risk. Second, in addition to the dose constraint for all exposure pathways, concentrations of radionuclides in groundwater should be limited in accordance with current standards for public drinking-water supplies unless compliance with drinking-water standards is not feasible.

The difference of opinion between EPA and the Nuclear Regulatory Commission about the adequacy of the Nuclear Regulatory Commission's cleanup standards for contaminated sites is strictly a matter of differences in policy judgments for risk management. Those judgments include the determination of a limit on acceptable risk and therefore dose for this exposure situation and the determination of whether separate requirements are needed for protection of groundwater resources.

Consistency of Radiation Guidelines

Many diverse guidelines have been developed by EPA and other organizations for TENORM and human-made radionuclides. The fundamental purpose of the guidelines is to limit risks to exposed individuals and populations. However, when the various guidelines are compared, the levels of acceptable risk corresponding to the numerical criteria in the guidelines appear to be inconsistent. The committee has considered the issue of consistency of radiation guidelines with regard to risk and offers the following observations.

First, although the desire for consistency of guidelines with regard to levels of acceptable risk is understandable, the committee has identified several important reasons why such a consistency should not be expected:

- Differences in statutory and judicial mandates for guidelines, especially the fundamental difference between a regulatory limit, as embodied in some guidelines, and a regulatory goal that can be relaxed on the basis of other considerations as embodied in other guidelines.

- Differences in the primary bases of guidelines, especially judgments about acceptable risk versus judgments about risks that are reasonably achievable.

- Differences in the applicability of guidelines, especially guidelines that apply to all sources of exposure combined versus guidelines that apply only to specific sources or practices, or to particular environmental media and comparisons of guidelines that apply to quite different sources or practices.

- Differences in the population groups of primary concern, especially individuals who receive the highest exposures versus whole populations.

- Differences in the considerations of natural background.

The committee believes that it is important to understand those factors when comparing different guidelines.

Second, the numerical criteria in the guidelines, whether they are in the form of limits or goals, do not appear to be the most important factor in determining acceptable risks to individuals or populations. Rather, without regard for the substantial differences in risks corresponding to the various guidelines and without regard for the different factors that result in these differences, as described above, the principle that exposures of individuals and populations should be maintained as low as reasonably achievable (ALARA), economic and social factors being taken into account, appears to be the most important factor in determining risks actually experienced for any exposure situation that is subject to regulatory control. Therefore, to the extent that the ALARA objective is applied consistently to all exposure situations, all guidelines would be consistent with regard to the risks actually achieved, even though the risks that are ALARA can depend significantly on the particular exposure situation.

Importance of Natural Background for Guidelines for TENORM

Regulation of TENORM is a unique problem among all radioactive materials in that the radionuclides of concern occur naturally in all environmental media. Therefore, guidelines for TENORM must correspond to levels of naturally occurring radionuclides in the environment at which it is practical to distinguish the radionuclides resulting from human activities from those in the undisturbed natural background. Furthermore, determinations of practical levels for identifying and controlling TENORM must take into account the variability of natural levels in different environmental media, as well as the average values.

Importance of Knowledge of Sources of TENORM

Sources of TENORM other than indoor radon result from a wide variety of human activities, and the physical and chemical properties of the radioactive materials that result from those activities vary widely. Furthermore, some sources are discrete and thus localized, whereas others are diffuse and very large in volume. Especially when guidelines for TENORM might be expressed in terms of concentrations in environmental media (such as soil), rather than dose or risk, development of the guidelines should take into account the properties of the various sources of concern. It would be inappropriate to apply a guideline developed for a particular exposure situation to other situations when there are important differences in the properties of the radioactive materials.

In addition, exposure-pathway analysis for TENORM generally should take into account the leachability, sorption, and biologic availability of the particular physical and chemical forms of the materials. In some cases, the characteristics of TENORM differ substantially from the characteristics of naturally occurring materials in their undisturbed state.

Differences in Approaches to Radiation Risk Assessment

As part of this study, the committee considered the approaches currently used by EPA in estimating cancer risks posed by radiation exposure, as documented in *Federal Guidance Report No. 13 (Part 1, interim version)*, in comparison with the approaches used by most other organizations whose guidelines were evaluated. Most other organizations estimate cancer risks on the basis of methods developed by ICRP—either the current methods represented in ICRP Publications 60 and 72 and supporting documents or, in the case of other federal agencies and state organizations, the methods represented in ICRP Publications 26 and 30, which have been superseded by Publications 60 and 72.

The issue of differences in approaches to risk assessment is of concern only for radionuclides other than radon because the risks posed by exposure to radon can be estimated, with some uncertainty, from epidemiologic data without the need to estimate the dose to radiosensitive tissues per unit exposure and the risk per unit dose for different types of radiation. In addition, for external exposure, the differences between risks estimated by EPA and the risks estimated by other organizations and based on ICRP methods generally are insignificant.

For internal exposure, EPA's approach to risk estimation differs from approaches based on current or outdated ICRP methods in three important respects.

- First, EPA's risk models take into account the age dependence of the absorbed dose rate in body tissues and the age dependence of the radiogenic risks, whereas risks calculated with ICRP methods are based on calculated committed effective doses or effective dose equivalents, which are not intended to provide accurate representations of cancer risks for individual organs and tissues of the body. The difference between the two approaches is particularly important for long-lived radionuclides with long retention times in the body.

- Second, EPA's assumptions about mortality for all cancers and for other competing causes of death, which are used in estimating radiogenic risks, are based on data on the U.S. population, which differ from the data used by ICRP, and there are other differences in the risk estimates for particular cancers.

- Third, the other federal agencies and state governments that normally estimate risks based on the outdated ICRP methods do not take into account age-specific dosimetric and biokinetic models and current models for the redistribution and retention of radioactive decay products in the body.

The differences between EPA's risk estimates for internal exposure and estimates obtained with ICRP methods are particularly important for the long-lived, alpha-emitting radionuclides found in TENORM (such as thorium). If EPA's risk estimates are compared with estimates based on the outdated ICRP methods normally used by other federal agencies and state governments, the differences are well in excess of a factor of 10 in some cases. In comparison with current ICRP methods based on age-specific committed effective doses, the differences can be as large as a factor of 5. EPA's risk estimates are lower in all cases.

The EPA uses its methodologically more rigorous approaches to risk assessment only in assessing risks for purposes of reaching decisions on rule-making, including decisions on the feasibility of establishing guidelines and the effects of alternative guidelines. However, when guidelines are expressed in terms of dose, as is often the case, EPA uses standard ICRP calculations of committed effective dose equivalents for adults, based on the methods and data in ICRP Publications 26 and 30, for purposes of demonstrating compliance with the guidelines to maintain a stable and uniform framework for the regulated community.

The committee generally supports the current EPA approaches to estimating risks posed by radiation exposure. They should be appropriate for the current US population, and the methods for estimating risks posed by internal exposure are methodologically more rigorous than those used by other organizations on the basis of current or outdated ICRP methods incorporating committed doses.

EPA's current methods of risk assessment as they differ from the methods used by most other organizations—especially for long-lived, alpha-emitting radionuclides—have not had a direct influence on the development of guidelines for TENORM expressed in terms of dose. That is mainly because guidelines expressed in terms of dose apply to both external and internal exposure, EPA's risk estimate for external exposure is nearly the same as

ICRP's, and external exposure often is important in scenarios for TENORM. Furthermore, policy judgments for risk management and a desire for consistency in regulation have been more important for the development of guidelines for TENORM than any differences in estimated risks based on different methods. However, if EPA chose to develop guidelines for TENORM in the form of concentrations in environmental media, its current methods for risk assessment could be used to derive the guidelines from an assumed limit on acceptable dose or risk. Such an approach could be suitable for TENORM, because only a few radionuclides are of concern. However, the analysis of exposure pathways and dose would need to account for the various physical and chemical forms of radionuclides that may be encountered in the environment.

Use of Linear, No-threshold Dose-Response Hypothesis

At the present time, there is considerable debate over the validity of the linear, no-threshold dose-response hypothesis for low levels of exposure. It remains as an assumption used in developing all radiation guidelines, including those for TENORM, in spite of the current debate over the validity of this hypothesis, including the possibilities that there is an effective threshold for radiation risks and that there are beneficial effects at low doses.

The committee does not have any new insights into the validity of the linear, no-threshold dose-response hypothesis. However, the committee understands that it is used because it represents a prudent approach to health protection of the public in the absence of definitive information on radiation risks at the dose levels of concern for routine exposures. The committee also notes that the central issue here is the risks due to incremental increases in dose above background, not the risks due to the incremental increases themselves.

Directions for Further Research on TENORM

The committee has noted a number of subjects on which additional scientific information would be beneficial in developing guidelines for TENORM.

Although models for exposure and dose assessment generally are well developed, the models for TENORM, especially the models for exposure assessment, possibly could be improved through validation of parameters. Much information on exposure-pathway models for naturally occurring radionuclides was obtained in studies on uranium mill tailings, but this information may be inappropriate for other exposure situations involving substantially different physical and chemical forms of radionuclides. Differences in chemical and physical forms of TENORM also could affect the estimates of dose from ingestion and inhalation of radionuclides.

There is need for improved methods for locating and measuring discrete and diffuse TENORM in the environment, especially if guidelines for TENORM correspond to levels in the environment that are only marginally above the levels of natural background.

Finally, given the importance of the linear, no-threshold dose-response hypothesis, an understanding of radiation carcinogenesis and the validity of the hypothesis remains an important scientific need for radiation protection, specifically for estimating the probabilities of adverse human health effects at the levels of natural background.

1
Introduction

The presence of naturally occurring radionuclides in soil, rock, water, and air along with cosmic radiation results in continuous and largely unavoidable radiation exposures of all humans. Exposures larger than these due to undisturbed natural background can result from human activities that move naturally occurring radionuclides from normally inaccessible locations to locations where humans are present or concentrate naturally occurring radionuclides. Examples of human activities that can increase exposures to naturally occurring radionuclides by relocation or concentration are mining and milling of mineral ores, extraction of petroleum products, use of groundwater for domestic purposes, and living in houses. The present US inventory of waste materials generated by these activities is in excess of 60 billion metric tons (EPA 1993b).

DEFINITION AND SCOPE OF STUDY

This study by the National Research Council's Committee on the Evaluation of EPA Guidelines for Exposures to Naturally Occurring Radioactive Materials (NORM) was initiated in response to a congressional directive that included the following:

> The Administrator of the Environmental Protection Agency (EPA) shall enter into an arrangement with the National Academy of Sciences to investigate and report on the scientific bases for the public recommendations of EPA with respect to indoor radon and other naturally occurring radioactive materials (NORM). The National Academy shall examine EPA's guidelines in light of the recommendations of the National Council on Radiation Protection and

Measurements, and other peer-reviewed research by the National Cancer Institute, the Centers for Disease Control, and others, on radon and NORM. The National Academy shall summarize the principal areas of agreement and disagreement among the above, and shall evaluate the scientific and technical basis for any differences that exist.

Exposures to naturally occurring radionuclides resulting from human activities that alter the natural environment can, to some degree, be subjected to regulatory control. The general concern of this study is the regulation of exposures to technologically enhanced naturally occurring radioactive materials by the US Environmental Protection Agency (EPA) and other regulatory and advisory organizations. The issues are complex and have been the focus of attention of radiation-protection authorities at the state and federal levels for over 2 decades (for example, CRCPD 1997; Bliss 1978; CRCPD 1978). Pursuant to the congressional mandate, the National Research Council's committee was charged with undertaking a study to compare EPA guidelines for controlling exposures of the public to technologically enhanced NORM with other regulatory and advisory organizations' guidelines. The concern expressed by Congress is that different organizations had arrived at different numerical values for guidelines that apply to essentially the same exposure situations and that the reasons for the differences, especially their scientific and technical bases, were not apparent.

Specifically, the committee was asked to address the following questions:

- Whether the differences among EPA and other guidelines for technologically enhanced NORM are based on scientific and technical information or on policy decisions related to risk management.

- If there are differences in the scientific and technical bases of the guidelines developed by EPA and other organizations, what the relative merit of the different bases is.

- Whether there is appropriate scientific information that has not been used in the development of contemporary risk analysis for NORM.

In regard to the third question, the committee was asked to evaluate whether there might be important differences in approaches to risk assessment between NORM and human-made radionuclides.

On the basis of the congressional mandate and the charge to the committee, the committee defined the scope of the study as follows:

- This study is concerned with guidelines for controlling exposures of the public to NORM, but it is not concerned with guidelines for NORM in occupational settings or with guidelines for human-made radionuclides in the workplace or the environment.

- This study is not concerned with issues of exposure during, risks posed by, or guidelines for transportation of NORM.

- This study is concerned with the extent to which the assumptions about radiation risks used by EPA in developing its guidelines for NORM differ from the assumptions used by other organizations in developing similar guidelines and with the relative merits of assumptions about radiation-related risks. The study is not concerned with other issues of risk assessment for NORM that may arise in attempting to estimate real risks posed by actual exposures of individuals or populations.

- This study does not include an independent evaluation of current information on radiation-related risks from epidemiologic studies. That kind of evaluation is the responsibility of such other authoritative organizations as the National Research Council's Committee on the Biological Effects of Ionizing Radiations (BEIR), the United Nations Scientific Committee on the Effects of Atomic Radiation (UNSCEAR), the National Council on Radiation Protection and Measurements (NCRP), and the International Commission on Radiological Protection (ICRP). In particular, the extensive information on the health risks from exposure to the important naturally occurring materials radium, thorium, uranium, and radon, based on human and animal studies, has been reviewed and evaluated, for example, by the BEIR IV Committee (National Research Council 1988) and by Stannard (1988), but is not evaluated in this study.

- This study is not concerned with issues of site-specific risk assessments for NORM. But some issues that are important to site-specific assessments are also important in the development of guidelines for NORM, such as the dependence of exposure pathways on the physical and chemical forms of radionuclides, and these issues are considered in this study.

INTRODUCTION

- This study is not concerned with evaluations of nonscientific issues that are important in the development of guidelines for NORM, including costs and policy judgments in risk management. But in evaluating whether differences among EPA and other guidelines for NORM are based on scientific and technical information, the committee found it necessary to consider policy judgments in risk management that have influenced the development of guidelines, even though the merit of any judgments is not evaluated in this study.

- This study is not concerned with exposures of, risks to, or guidelines for biota other than humans.

THE RADIOACTIVE MATERIALS OF CONCERN

The committee has used the term *technologically enhanced naturally occurring radioactive materials* (TENORM) to refer to the materials of concern in this study. The committee defines this term as follows:

Technologically enhanced naturally occurring radioactive materials are any naturally occurring radioactive materials not subject to regulation under the Atomic Energy Act whose radionuclide concentrations or potential for human exposure have been increased above levels encountered in the natural state by human activities.

The exclusion of NORM that are regulated under the Atomic Energy Act from the definition of TENORM is important. This study is not concerned with evaluating guidelines developed by EPA or other organizations that apply to NORM associated with the production and use of nuclear fuels such as uranium and thorium mill tailings, any naturally occurring radionuclides released to the environment during operations of nuclear fuel-cycle facilities, or natural uranium or thorium in the form of source material. However, the committee has considered EPA guidelines for NORM arising from the nuclear fuel cycle because these guidelines have provided important precedents in establishing guidelines for similar non-fuel-cycle radioactive materials of concern to this study.

The distinction between NORM associated with the nuclear fuel cycle, which are not of concern in this study, and TENORM as defined above is rooted in the definitions of source, special nuclear, and byproduct materials as used in the Atomic Energy Act (see appendix A) and the authority given to the US Atomic Energy Commission (and, later, the US Nuclear Regulatory Commission) to regulate radioactive materials defined in the act but not any other radioactive materials. Before the establishment of EPA in 1970 and the promulgation of environmental laws concerned primarily with hazardous materials other than radionuclides, there was no legal authority for federal regulation of TENORM, and the regulation of radioactive materials from the nuclear fuel cycle has proceeded largely independently of the regulation of TENORM, with considerably greater attention given to fuel-cycle materials.

Thus, the concern about guidelines for TENORM that led to this study arises from the definitions of radioactive materials in the Atomic Energy Act and the much greater attention that has been given to the regulation of radioactive materials from the nuclear fuel cycle. However, the distinction between naturally occurring radionuclides associated with the nuclear fuel cycle and naturally occurring radionuclides associated with other activities is artificial with regard to protection of human health and the environment in that the risks posed by a given radiation exposure do not depend on the source of the radioactive material.

APPROACH TO THE STUDY

In evaluating guidelines for TENORM developed by EPA and other organizations in response to the charge described above, the committee has considered guidelines for indoor radon separately from guidelines for any other TENORM. The distinction between radon and any other radionuclides in the context of developing guidelines for radiation exposure was incorporated in the earliest standards for limiting internal exposures of workers (Advisory Committee on X-ray and Radium Protection 1941), and has been maintained ever since in developing guidelines for workers and the public.

There are two important scientific and technical reasons for the distinction between radon and other radionuclides. First, radon is an inert gas; it is the only naturally occurring radionuclide in this form. Its emanation from radium-bearing soil, rock, and building materials results in substantial exposures in indoor environments, and it is unique among the radionuclides with regard to the importance of this exposure pathway. Indeed, indoor radon is the most important source of radiation exposure of the public.

Second, and more important, the relationship between exposure to short-lived decay products of radon in air and the risk of lung cancer can be estimated, with some uncertainty, from epidemiologic studies in various groups of miners. Thus, the risk posed by exposure to indoor radon can be estimated without the need to develop models for estimating doses to radiosensitive tissues of the lung from irradiation by alpha particles after inhalation and without the need to invoke assumptions about the risk per unit dose of alpha particles. Radon is unique among the radionuclides in this regard; for no other radionuclides can estimates of cancer risk posed by internal exposure be obtained without estimating the dose per unit intake and the risk per unit dose. Estimates of risk associated with external exposure to any radionuclide also must be based on estimates of dose and risk per unit dose.

Evaluation of Guidelines for Indoor Radon

The approach in this study to evaluating guidelines for indoor radon developed by EPA and other organizations is relatively straightforward because the guidelines for this exposure situation are well defined. The main task of the committee was to evaluate whether the differences among the various guidelines have a scientific and technical basis.

Evaluation of Guidelines for TENORM Other Than Indoor Radon

The committee's task in evaluating guidelines for TENORM other than indoor radon was more difficult than the task in evaluating guidelines for indoor radon, in part because of the variety of exposure situations of concern, including releases from controlled sources, waste management and disposal, and remediation of environmental contamination. In addition, for the following reasons, it was not a simple matter for the committee to define the current EPA guidelines for these materials.

First, some guidelines are outdated and seem not to represent EPA's current views on suitable approaches to regulating TENORM. In some cases, proposals for revising guidelines have been published in the *Federal Register* to provide opportunity for public comment, but revised guidelines have not been issued in final form. Thus, it was not clear at the time of this study whether the existing but outdated guidelines would be replaced or, if so, what the new guidelines would be. An important example is EPA's proposed federal guidance on radiation protection of the public, which would replace guidance developed by the Federal Radiation Council (FRC) nearly 40 years ago. The proposed revision of the federal guidance contains a dose limit for all controlled sources of exposure combined (including human-made radionuclides and TENORM other than indoor radon) that is one fifth the dose limit in the FRC guidance. The question of which guidance represents EPA's current views on radiation protection of the public is obviously important in comparing EPA guidelines for TENORM with guidelines developed by other organizations.

Second, some EPA guidelines for specific exposure situations involving TENORM are in the form of legally enforceable regulations that were published in the *Federal Register* and took public comment into account, but EPA's preferred approach to regulating TENORM in other situations is indicated only by policy statements that have not been subjected to public comment. Important examples are EPA's strategy for protecting groundwater resources and various EPA directives on the interpretation of requirements for remediation of contaminated sites under the Comprehensive Environmental Response, Compensation, and Liability Act (CERCLA). The committee was faced with decisions about the weight to be given EPA guidelines in the form of policy statements relative to those in the form of legally enforceable regulations.

Third, for some important exposure situations, such as management and disposal of waste that contains TENORM except waste arising from treatment of drinking water, EPA has not published any guidelines, and the committee had to infer EPA's preferred approach to regulation on the basis of existing or proposed guidelines for similar exposure situations.

Fourth, no EPA regulation or set of regulations applies to all potentially important sources of exposure to TENORM other than indoor radon.

For all those reasons, the committee had to use judgment in selecting EPA guidelines for TENORM other than indoor radon to be given the greatest emphasis in comparisons with guidelines developed by other organizations. In general, the committee has emphasized more recent guidelines without regard for other considerations, such as whether they have been issued in proposed or final form and whether they are legally binding regulations or only policy statements. However, in an effort to provide a reasonably complete picture of EPA's current policies, the committee has attempted to discuss all EPA guidelines of any kind that are relevant to the regulation of TENORM.

Other Considerations

The committee has interpreted its charge to review and evaluate guidelines for TENORM quite broadly. It has endeavored to go beyond simple recitations of facts and figures for the guidelines of concern. There are many and diverse guidelines, especially for TENORM other than indoor radon, and many of them seem inconsistent with regard to the acceptable health risks for the public. The committee believes it important to describe the bases for the various guidelines, how they should be interpreted, and how they fit within an overall framework for radiation protection of the public that embodies only a few basic principles. Guidelines are important individually, but it also is important to understand how they are related and how they are consistent with one another.

In protection of public health, there have been different uses among various federal agencies and other organizations of such terms as acceptable and unacceptable to describe different levels of risk, and some organizations also have used <u>tolerable</u> and <u>intolerable</u>. In discussing any particular guideline for TENORM, the committee has endeavored to use such terms as they were used by the organization whose guideline is under consideration. However, because the terms, especially *acceptable* and *unacceptable*, sometimes have been used in ways that do not conform to the public's general understanding of their meaning and import, the committee has commented on their possible misinterpretations in discussing particular guidelines.

Although the purpose of any guideline is to limit health risks, guidelines themselves usually are expressed in terms of exposure or dose, rather than directly in terms of risk. The earliest radiation standards often were expressed in terms of limits on dose to the whole body or the critical organ, which usually is the organ that receives the highest dose. Later standards have been expressed in terms of the effective dose equivalent or effective dose, as defined by the ICRP. Those quantities are weighted sums of doses to several organs and tissues that are intended to be proportional to risk posed by any uniform or nonuniform irradiation of the whole body, the latter often occurring as a result of intakes of radionuclides by ingestion or inhalation. An understanding of the various concepts and terms for radioactivity and radiation dose is important for understanding risk, and a glossary of radiation quantities and units is given in the appendix.

STRUCTURE OF THE REPORT

The remaining chapters in this report are organized into three groups, as follows.

Chapters 2-6 provide background information that is important for understanding and evaluating the various guidelines for TENORM: information on characteristics of naturally occurring radionuclides and natural background radiation, important sources of TENORM, the role of exposure pathway and dose or risk assessments in providing a technical basis for radiation standards, the basic judgments involved in developing radiation standards for any exposure situation, and the responsibilities of the regulatory and advisory organizations whose guidelines for TENORM have been considered in this study.

Chapters 7-9 present detailed information on the guidelines for TENORM considered in this study. Chapter 7 reviews the existing or proposed guidelines developed by EPA. It also discusses health risks to the public corresponding to the different guidelines, the issue of consistency of guidelines with regard to limits on risk, the relationship between the guidelines and risks experienced in actual exposure situations, and the importance of the objective that exposures be as low as reasonably achievable (ALARA) in determining actual risks. Chapter 8 reviews the existing guidelines for indoor radon developed by EPA and other organizations and information on radon-related risks assumed by the various organizations. Chapter 9 reviews the existing guidelines for TENORM other than indoor radon developed by organizations other than EPA. It also discusses the issue of transferability from one exposure situation to another of guidelines in the form of limits on concentrations of radionuclides in environmental media.

Chapters 10-12 focus on the charge to the committee on the basis of information presented in the preceding chapters. Chapter 10 presents summaries and comparisons of the various guidelines for TENORM reviewed in Chapters 7-9. Guidelines developed by EPA for indoor radon and for TENORM other than indoor radon are compared with guidelines developed by other organizations, and the committee's views on the reasons for the differences are presented. Chapter 11 mainly presents summary discussions on the question of whether the differences between EPA and other guidelines for TENORM have a scientific and technical basis, specific ways in which the technical approaches to risk assessment of radionuclides currently used by the EPA and other organizations differ and whether the differences have been important in developing guidelines for TENORM, and specific ways in which the differences between EPA and other guidelines for TENORM are based on policies related to risk management, rather than scientific and technical issues. Chapter 12 presents some summary conclusions and recommendations developed by the committee during the course of this study.

Systeme International (SI) units are used throughout this report. However, the conventional units for such quantities as activity, exposure, and dose are given in parentheses in many cases because most regulations in the United States use the conventional units. The relationships between SI and conventional units are given in the appendix.

2
Natural Radioactivity and Radiation

This chapter describes the behavior of selected natural radionuclides in the environment, the sources and variability of natural radiation, and the doses received by humans. Its purpose is to provide background information for discussions of the mechanisms by which exposures to natural sources can be increased by technologic activities, that is, can become exposures to TENORM. A more detailed account of natural radiation can be found in Eisenbud and Gesell (1997), which was used as a guide to prepare parts of this chapter.

Natural radiation comprises cosmic radiation and the radiation arising from the decay of naturally occurring radionuclides. The natural radionuclides include the primordial radioactive elements in the earth's crust, their radioactive decay products, and radionuclides produced by cosmic-radiation interactions. Primordial radionuclides have half-lives comparable with the age of the earth. Cosmogenic radionuclides are produced continuously by bombardment of stable nuclides by cosmic rays, primarily in the atmosphere.

Humans are exposed to natural radiation from external sources, which include radionuclides in the earth and cosmic radiation, and by internal radiation from radionuclides incorporated into the body. The main routes of radionuclide intake are ingestion of food and water and inhalation. A particular category of exposure to internal radiation, in which the bronchial epithelium is irradiated by alpha particles from the short-lived progeny of radon, constitutes a major fraction of the exposure from natural sources.

In most places on the earth, natural radiation from external sources varies within about a factor of 4; but in some localities, the variation is greater because of abnormally high or low soil concentrations of radioactive minerals. Cosmic radiation alone varies by about a factor of 2 over the range of elevation that encompasses most of the world's population (0-2,000 m) and to a much smaller degree with latitude because of the variation in the earth's magnetic field. Particularly high concentrations of radioactive minerals in soil have been reported in Brazil, India, and China. Variations of radon concentrations in buildings are responsible for the largest variations in doses received by the public from natural internal sources.

NATURALLY OCCURRING RADIONUCLIDES

The origin of the primordial natural radionuclides of the earth is associated with the phenomenon of nucleosynthesis in stars (Fowler 1967). The fact that the uranium, thorium, and actinium decay chains are found in nature is directly related to the very long half-lives of the parents of these chains. The absence of the neptunium decay chain is due to the lack of sufficiently long-lived members of this chain; complete decay of the parent radionuclides and their progeny has already occurred. Naturally occurring radionuclides with long half-lives that are not members of decay chains also exist in relatively high isotopic abundance.

For purposes of discussion, the naturally occurring radionuclides are divided into those which occur singly (tables 2.1 and 2.2) and those which are components of three chains of radioactive elements. The uranium chain (table 2.3) originates with ^{238}U; the thorium chain (table 2.4), with ^{232}Th; and the actinium chain (table 2.5), with ^{235}U. Each table shows the nuclide, half-life, and principal radiations associated with each important branch of the chain. Minor branches, (less than 1%) and natural fission[2] are not listed, nor do they make any important contribution to the radiation dose from these chains. Tables 2.1 and 2.2 also show typical concentrations in various environmental media.

[2]In nature, ^{235}U and a few other nuclides of uranium and thorium undergo fission spontaneously or as a result of interactions with neutrons that originate in cosmic rays or other natural sources. The half-life of ^{235}U owing to spontaneous fission is 10^{15}-10^{16} y, so decay by this process is at a rate less than 10^{-7} of that due to alpha-particle emission.

Table 2.1 Radionuclides Induced in Earth's Atmosphere by Cosmic Rays[a]

Radio-nuclide	Half-life	Major Radiations	Target Nuclides	Typical Concentrations, Bq/kg		
				Air (troposphere)	Rain Water	Ocean Water
^{10}Be	1,600,000 y	β	N, O	--	--	2×10^{-8}
^{26}Al	716,000 y	γ	Ar	--	--	2×10^{-10}
^{36}Cl	300,000 y	β	Ar	--	--	1×10^{-5}
^{81}Kr	229,000 y	K x rays	Kr	--	--	--
^{14}C	5730 y	β	N, O	--	--	5×10^{-3}
^{32}Si	172 y	β	Ar	--	--	4×10^{-7}
^{39}Ar	269 y	β	Ar	--	--	6×10^{-8}
^{3}H	12.33 y	β	N, O	1.2×10^{-3}	--	7×10^{-4}
^{22}Na	2.60 y	β^{+}	Ar	1×10^{-6}	2.8×10^{-4}	--
^{35}S	87.51 d	β	Ar	1.3×10^{-4}	$7.7 \times 10^{-3} - 107 \times 10^{-3}$	--
^{7}Be	53.29 d	γ	N, O	0.01	0.66	--
^{37}Ar	35.0 d	K x rays	Ar	3.5×10^{-5}	--	--
^{33}P	25.3 d	β	Ar	1.3×10^{-3}	--	--
^{32}P	14.26 d	β	Ar	2.3×10^{-4}	--	--
^{28}Mg	20.91 h	β	Ar	--	--	--
^{24}Na	14.96 h	β	Ar	--	$3.0 \times 10^{-3} - 5.9 \times 10^{-3}$	--
^{38}S	2.84 h	β	Ar	--	$6.6 \times 10^{-2} - 21.8 \times 10^{-2}$	--

[a] Adapted from NCRP (1987a) and NuDat online database maintained by Brookhaven National Laboratory, September 9, 1997.

Table 2.1 (continued)

Radio-nuclide	Half-life	Major Radiations	Target Nuclides	Typical Concentrations, Bq/kg		
				Air (troposphere)	Rain Water	Ocean Water
^{31}Si	2.62 h	β	Ar	--	--	--
^{18}F	1.83 h	β^+	Ar	--	--	--
^{39}Cl	55.6 m	β	Ar	--	1.7×10^{-1} - 8.3×10^{-1}	--
^{38}Cl	37.24 m	β	Ar	--	1.5×10^{-1} - 25×10^{-1}	--
^{34m}Cl	32.0 m	β^+	Ar	--	--	--

Table 2.2. Nonchain Primordial Radionuclides[a]

Radionuclide	Half-life, y	Major Radiations	Typical Crustal Concentration, Bq/kg
^{40}K	1.28×10^9	β, γ	630
^{50}V	1.4×10^{17}	γ	2×10^{-5}
^{87}Rb	4.75×10^{10}	β	70
^{113}Cd	9×10^{15}	β	$<2 \times 10^{-6}$
^{115}In	6×10^{14}	β	2×10^{-5}
^{123}Te	1.24×10^{13}	x rays	2×10^{-7}
^{138}La	1.05×10^{11}	β, γ	2×10^{-2}
^{142}Ce	$>5 \times 10^{16}$	β	$<1 \times 10^{-5}$
^{144}Nd	2.29×10^{15}	α	3×10^{-4}
^{147}Sm	1.06×10^{11}	α	0.7
^{152}Gd	1.08×10^{14}	α	7×10^{-6}
^{174}Hf	2.0×10^{15}	α	2×10^{-7}
^{176}Lu	3.73×10^{10}	β, γ	0.04
^{187}Re	4.3×10^{10}	β	1×10^{-3}
^{190}Pt	6.5×10^{11}	α	7×10^{-8}

[a] Adapted from NCRP (1987a) and NuDat online database maintained by Brookhaven National Laboratory, September 9, 1997.

Table 2.3 Uranium-238 Chain[a]

Nuclide	Historical Name	Half-life	Major Radiations
^{238}U	Uranium I	4.47×10^9 y	α, < 1% γ
^{234}Th	Uranium X_1	24.1 d	β,
^{234m}Pa	Uranium X_2	1.17 m	β, < 1% γ
^{234}U	Uranium II	2.46×10^5 y	α, < 1% γ
^{230}Th	Ionium	7.54×10^4 y	α, < 1% γ
^{226}Ra	Radium	1600 y	α, γ
^{222}Rn	Emanation	3.82 d	α, < 1% γ
^{218}Po	Radium A	3.10 m	α, < 1% γ
^{214}Pb	Radium B	26.8 m	β, γ
^{214}Bi	Radium C	19.9 m	β, γ
^{214}Po	Radium C'	164.3 μs	α, < 1% γ
^{210}Pb	Radium D	22.3 y	β, γ
^{210}Bi	Radium E	5.01 d	β
^{210}Po	Radium F	138.4 d	α, < 1% γ
^{206}Pb	Radium G	Stable	None

[a]Data from NuDat online database maintained by Brookhaven National Laboratory, September 9, 1997. Minor branches, <1%, not shown.

Table 2.4 Thorium-232 Chain[a]

Nuclide	Historical Name	Half-life	Major Radiations
^{232}Th	Thorium	1.41×10^{10} y	α, <1%γ
^{228}Ra	Mesothorium I	5.75 y	β, <1% γ
^{228}Ac	Mesothorium II	6.15 h	β, γ
^{228}Th	Radiothorium	1.91 y	α, γ
^{224}Ra	Thorium X	3.66 d	α, γ
^{220}Rn	Emanation	55.6 s	α, <1% γ
^{216}Po	Thorium A	0.145 s	α, <1% γ
^{212}Pb	Thorium B	10.64 h	β, γ
^{212}Bi	Thorium C	1.01 h	α, γ
^{212}Po (64%) ^{208}Tl (36%)	Thorium C' / Thorium C''	0.300 ms / 3.05 m	α / β, γ
^{208}Pb	Thorium D	Stable	None

[a] Data from the NuDat online database maintained by Brookhaven National Laboratory, September 9, 1997. Minor branches, <1%, not shown.

Table 2.5 Uranium-235 (Actinium) Chain[a]

Nuclide	Historical Name	Half-life	Major Radiations
^{235}U	Actinouranuim	7.04×10^8 y	α, γ
^{231}Th	Uranium Y	1.06 d	β, γ
^{231}Pa	Protoactinium	3.28×10^4 y	α, γ
^{227}Ac	Actinium	21.77 y	β, < 1% γ
^{227}Th (98.62%) / ^{223}Fr (1.38%)	Radioactinium / Actinium K	18.72 d / 22.0 m	α, γ / β, γ
^{223}Ra	Actinium X	11.44 d	α, γ
^{219}Rn	Actinon	3.96 s	α, γ
^{215}Po	Actinium A	1.78 ms	α, < 1% γ
^{211}Pb	Actinium B	36.1 m	β, γ
^{211}Bi	Actinium C	2.14 m	α, γ
^{207}Tl	Actinium C'	4.77 m	β, < 1% γ
^{207}Pb	Actinium D	Stable	None

[a]Data from NuDat online database maintained by Brookhaven National Laboratory, September 9, 1997. Minor branches, <1%, not shown.

The three chains of radioactive elements and the long-lived primordial nuclide potassium-40 account for much of the external background radiation dose from radionuclides to which humans are exposed. Of the 22 nuclides identified as cosmogenic (table 2.1) only two, carbon-14 and tritium (^{3}H), are of any consequence from the perspective of dose to humans. Only two of the 15 nonchain primordial nuclides, ^{40}K and rubidium-87, are of particular interest (table 2.2).

Uranium and thorium can be concentrated in rocks by igneous and sedimentary processes (Bliss 1978). Where uranium and thorium concentrations are high enough, rocks constitute ores to industrial societies. In the western United States, uranium ores have been extensively mined and milled to produce nuclear fuels.

The biogeochemical behavior of a radionuclide in a given decay chain can be expected to vary with atomic number (that is, the element). For example, in the uranium decay chain, isotopes of uranium, thorium, radium, radon, and other elements occur. Chemically they range from an inert gas (radon) to a readily sorbed, tetravalent cation (thorium). Those properties determine the fate of the radionuclides in fuel and mineral processing, their transport in soil or surface disposal environments, and ultimately their biologic availability and uptake; a knowledge of their behavior is essential for defining source terms and assessing doses.

Regulations for controlling exposure of the public to radionuclides are often dose-based. Because the doses result from interaction of humans with radionuclides contained in environmental media—air, water, soil, and biota—a knowledge of the behavior of naturally occurring radionuclides in these media is needed (Landa 1980). It is important to know:

- The different mobilities of the various radionuclides in the decay chains.

- How technologic processes have changed the physical and chemical form of radionuclides and the release rates of radionuclides to the various media.

- How naturally occurring radioactive materials evolve with time (weathering reactions).

- The concentrations and physical and chemical forms of the radionuclides.

The following sections discuss the naturally occurring radionuclides that are potentially important contributors to human exposure to TENORM.

Other natural radionuclides that are contributors to background radiation dose but not necessarily to exposure to TENORM are discussed for completeness, but in less detail.

Uranium

The primordial uranium found ubiquitously in nature consists of two isotopes with mass numbers of 235 and 238. In the earth's crust, ^{238}U constitutes 99.27% of the uranium by mass, and ^{235}U, the parent isotope of the actinium chain, 0.72%. ^{234}U, a shorter-lived member of the ^{238}U chain, is usually in radioactive equilibrium or near-equilibrium with the parent isotope.

Geochemistry Oxidation-reduction processes play a major role in the occurrence and behavior of uranium in aqueous environments. The dominant uranium valence states that are stable in geologic environments are the uranous (U^{4+}) and uranyl (U^{6+}) states, the former being far less soluble. Uranium transport generally occurs in oxidizing surface water and groundwater as the uranyl ion, UO_2^{2+}, or as uranyl fluoride, phosphate, or carbonate complexes. UO_2^{2+} and uranyl fluoride complexes dominate in oxidizing, acidic waters, whereas the phosphate and carbonate complexes dominate in near-neutral and alkaline oxidizing waters, respectively. Hydroxyl, silicate, organic, and sulfate complexes might also be important, the sulfate complex being important especially in mining and milling operations that use sulfuric acid as a leaching agent. Maximum sorption of uranyl ions on natural materials (organic matter; iron, manganese and titanium oxyhydroxides; zeolites, and clays) occurs at pH 5.0-8.5.

The sorption of uranyl ions by such natural media appears to be reversible; for uranium to be "fixed" and thereby accumulate, it requires reduction to U^{4+} by the substrate or by a mobile phase, such as H_2S.

Occurrence and Doses Uranium is found in all rocks and soils. Typical concentrations in the more prevalent types of rock and average concentrations in the earth's crust and in soil are listed in table 2.6. In the common rock types, the uranium concentrations range from 0.5 to 4.7 ppm, corresponding to activity concentrations for ^{238}U of 7-60 Bq/kg (0.2-1.6 pCi/g). The overall effect of soil development results in an average soil concentration of uranium less than the average rock concentration. Some ores mined and processed for nonradioactive materials can produce residues with elevated concentrations of radionuclides. A well-known example is phosphorus ore, which contains uranium at up to 120 ppm and has also been used as a commercial source of uranium (NCRP 1993b). Natural materials that contain uranium at over 500 ppm are considered to be uranium ores.

Uranium also occurs in air, water, and food and so is present in human tissues. The average annual intake of uranium from all dietary sources is about

Table 2.6 Ranges[a] and averages of concentrations of ^{40}K, ^{232}Th, and ^{238}U in Typical Rocks and Soils[b]

Material	^{40}K		^{232}Th		^{238}U	
	% total K	Bq/kg	ppm	Bq/kg	ppm	Bq/kg
Igneous rocks						
Basalt (crustal)	0.8	300	3-4	10-15	0.5-1	7-10
Mafic	1.1	300	2.7	10	0.9	10
Salic	4.5	1400	20	80	4.7	60
Granite(crustal)	>4	>1000	17	70	3	40
Sedimentary rocks						
Shale	2.7	800	12	50	3.7	40
Sandstones						
Clean quartz	<1	<300	<2	<8	<1	<10
Dirty quartz	2?	400?	3-6?	10-25?	2-3?	40?
Arkose	2-3	600-900	2?	<8	1-2?	10-25?
Beach sands	<1	<300	6	25	3	40
Carbonate rocks	0.3	70	2	8	2	25
All rock[a]	0.3-4.5	70-1400	2-20	7-80	0.5-4.7	7-60
Continental crust	2.8	850	10.7	44	2.8	36
Soil	1.5	400	9	37	1.8	22

[a] Examples of materials outside ranges can be found, but quantities are relatively small.
[b] Adapted from NCRP (1987a).

13 Bq (350 pCi) (NCRP 1987b). The intake of uranium from tap water can be a small or large fraction of the total intake depending on concentrations in local water supplies (Hess and others 1985). In the United States, the typical concentration of uranium in skeleton (wet weight) is about 8 mBq/kg (0.2 pCi/kg) (NCRP 1987b). Lung, kidney, and bone receive the highest annual doses of radiation from uranium, estimated at 11, 9.2, and 6.4 µSv (1.1, 0.92, and 0.64 mrem), respectively, for US residents. The decay products of uranium, particularly radium and its decay products, are more important than uranium itself with respect to dose to humans from both external and internal exposures (NCRP 1987a).

Radium-226

Radium-226 and its decay products, members of the uranium chain, are responsible for a major fraction of the internal dose received by humans from the naturally occurring radionuclides (IAEA 1990). ^{226}Ra is an alpha-particle emitter that decays, with a half-life of 1600 y, to radon-222, which has a half-life of 3.82 d (table 2.3). The decay of ^{222}Rn is followed by the successive disintegration of a number of short-lived alpha-particle- and beta-particle-emitting progeny. After six decay steps, in which radionuclides that range in half-life from 1.6×10^{-4} to 26.8 min are produced, ^{210}Pb is produced; it has a half-life of 22.3 y. This nuclide decays through ^{210}Bi to produce ^{210}Po, which decays by alpha-particle emission to stable ^{206}Pb. Radium itself adds little to the gamma-ray activity of the environment, but it does so indirectly through its gamma- ray-emitting decay products.

Geochemistry Radium exhibits only the +2 oxidation state in solution, and its chemistry resembles that of barium. Radium forms water-soluble chloride, bromide, and nitrate salts. The phosphate, carbonate, selenate, fluoride, and oxalate salts of radium are slightly soluble in water, whereas radium sulfate is relatively insoluble in water ($K_{sp} = 4.25 \times 10^{-11}$ at 20° C). Radium in uranium ore is only slightly soluble in H_2SO_4 but is highly soluble in HCl and HNO_3, presumably because of the greater solubility of $RaCl_2$ and $Ra(NO_3)_2$ than of $RaSO_4$.

The hydrated ion of radium is the smallest in the alkaline earth series, so it would tend to be preferentially retained by ion exchange. In alkaline solutions, anionic complexes of radium with organic ligands, such ethylenediamine tetraacetic acid (EDTA) and citric acid, are known to occur. Means and others (1978) suggest that EDTA mobilization might be responsible for elevated concentrations of radium seen in water and soil sampled around a radioactive-waste disposal trench at the Oak Ridge National Laboratory burial ground.

Radium does not form discrete minerals but can coprecipitate with many minerals, including calcium carbonate, hydrous ferric oxides, and barite ($BaSO_4$). Radium can be sorbed by clay minerals, colloidal silicic acid, manganese oxides, and organic matter. Although radium (unlike uranium) has only a single valence state, the dissolution or precipitation of sorbing phases, such as barite and ferric hydrous oxides, under changing oxidation-reduction conditions can influence its mobility. Groundwaters low in sulfate but high in ionic strength, calcium, and barium are conducive to the transport of radium.

Leaching data suggest that uranium mill tailings in the environment can constitute a long-term source of radium contamination of surface water and groundwaters that are in contact with them. The same is probably true of other NORM wastes in which ^{226}Ra is associated with sparingly soluble minerals, such as $BaSO_4$.

Occurrence and Doses ^{226}Ra is present in all rocks and soils in variable amounts. In nature, ^{226}Ra is generally in rough equilibrium with ^{238}U, so the concentrations compiled for ^{238}U in table 2.6 can be taken as a good guide to the expected range for ^{226}Ra. The radium contents of soils can show considerable spatial variability, both locally and regionally. These are the result of differences in parent materials and in soil-forming factors such as climate and weathering time. Soil-development processes can lead to substantial redistribution of macro-constituents, such as iron, and of trace elements and radionuclides, such as radium, in the soil profile, thereby introducing variations in distribution with depth, as well as location. The distribution of radium in uncontaminated, surface soils of the United States was investigated on a statewide-scale by Myrick and others (1981) in a study done in support of Department of Energy (DOE) remedial action programs dealing with fuel-cycle NORM. Individual ^{226}Ra measurements ranged from about 8.5 to 160 mBq/g (0.23 to 4.2 pCi/g). The state average ^{226}Ra measurements ranged from about 24 mBq/g (0.65 pCi/g) in Alaska to 56 mBq/g (1.5 pCi/g) in Kentucky, Nevada, New Mexico, and Ohio. Relative arithmetic standard deviations for the state averages ranged from 12 to 158%. The areal and cross-sectional variations that one might expect to see on smaller scales are exemplified in data presented by Meriwether and others (1995) and Van den Bygaart and Protz (1995), which show two-fold differences in ^{226}Ra concentration between surface horizons at different sampling sites and between surface and subsurface horizons at a given site. Spatial variability and other issues associated with soil sampling at sites that are potentially contaminated with radioactivity are discussed in detail in the *Multi-Agency Radiation Survey and Site Investigation Manual* (Nuclear Regulatory Commission/EPA 1996).

The radium content of surface waters (4-19 Bq/m^3, 0.1-0.5 pCi/L) is lower than that of most groundwaters (Hess and others 1985). Surveys of water supplies in many states (Cothern and Lappenbusch 1984) showed that the

Environmental Protection Agency (EPA) limit for total radium of 0.2 kBq/m^3 (5 pCi/L) was exceeded in many communities that obtain water from groundwater, including communities of about 600,000 in Illinois, Iowa, Missouri, and Wisconsin. About 75% of the supplies that exceeded 0.2 kBq/m^3 (5 pCi/L) were in two areas of the United States: the Piedmont and coastal plain areas of the Middle Atlantic states, and the north central states of Minnesota, Iowa, Illinois, Missouri, and Wisconsin. The concentration of ^{226}Ra was in some cases as high as 0.93 kBq/m^3 (25 pCi/L), with ^{228}Ra concentrations up to about 0.63 kBq/m^3 (17 pCi/L).

EPA (1991a) later conducted a random survey (stratified by system size) of radionuclides in 1,000 drinking-water supply systems that obtain water from ground water. For ^{226}Ra, 3.4 million persons were probably exposed to over 0.2 kBq/m^3 (5 pCi/L), and 890,000 to over 0.74 kBq/m^3 (20 pCi/L). The corresponding numbers are 1.3 million and 164,000 for ^{228}Ra. Persons consuming water that contains ^{226}Ra at 0.2 kBq/m^3 (5 pCi/L) at 2 L/d would receive an annual effective dose equivalent of about 50 µSv (5 mrem).

Radium is chemically similar to calcium and is absorbed from the soil by plants and passed up the food chain to humans. Because the radium in food originates in soil and the radium content of soil is variable, the radium content of foods varies. In addition, it is reasonable to expect that such chemical factors as the amount of exchangeable calcium in the soil will determine the rate at which radium is absorbed by plants. From radiochemical analyses of food, Fisenne and Keller (1970) determined the daily ^{226}Ra intake by inhabitants of New York City and San Francisco at 0.07 and 0.03 Bq (1.7 and 0.8 pCi), respectively. That difference is not reflected in the difference in ^{226}Ra content of human bone between the two cities (Fisenne and others 1981), which suggests an uncertainty of at least a factor of 2 in the relationship between intake and body burden. There is, however, an association between ^{226}Ra concentration in bone and the ^{226}Ra concentration in drinking water in the midwestern United States (NCRP 1987a). The National Council on Radiation Protection and Measurements (NCRP 1984c) estimates an average dietary intake of 0.05 Bq/d (1.3 pCi/d). Worldwide, the ^{226}Ra content of adult skeletons ranges from about 0.3 to 3.7 Bq (8 to 100 pCi), and the population-weighted average skeletal content is 0.85 Bq (23 pCi) (NCRP 1984c), which corresponds to annual equivalent doses of 170 µSv (17 mrem) to cortical and trabecular bone, 90 µSv (9 mrem) to the bone lining cells, 15 µSv (1.5 mrem) to the red marrow, and 3 µSv (0.3 mrem) to soft tissues.

Thorium

The only primordial isotope of thorium is thorium-232. Like uranium, it is ubiquitous in nature. Shorter-lived isotopes of thorium occur in all three of the natural decay chains.

Geochemistry In aqueous systems, only the Th^{4+} oxidation state is known to exist. Th^{4+} undergoes hydrolysis in aqueous solutions above pH 2-3 and is subject to extensive sorption by clay minerals and humic acid at near-neutral pH. At near-neutral pH and in alkaline soils, precipitation of thorium as a highly insoluble hydrated oxide phase and coprecipitation with hydrated ferric oxides can, with sorption reactions, be important mechanisms for the removal of thorium from solution. Because of sorption and precipitation reactions and the low solution rate of thorium-bearing minerals, thorium concentrations in natural waters are generally low.

At low pH, such as in an acid-leach uranium mill, thorium becomes more soluble. Acid-leach milling might dissolve 30-90% of the thorium in the ore. Acidic effluents (pH 2.5) from uranium mills in the Grants Mineral Belt of New Mexico contain ^{230}Th at 5.6-6.3 MBq/m^3 (150,000-170,000 pCi/L). The solubilized thorium can be precipitated if the acidic effluent is neutralized by contact with natural media or by process additions of limestone to the waste solutions. The high inventory of soluble ^{230}Th in such an effluent made it the radionuclide of greatest mobility when a dam at a New Mexico uranium mill failed in 1979, sending effluent down an arroyo (Weimer and others 1981). Similarly, under acidic conditions at some uranium mills, ^{230}Th has been shown to have migrated considerably deeper into the subsoil than ^{226}Ra (DOE 1993b).

Occurrence and Doses Typical concentrations of ^{232}Th in the more prevalent rock classes, the crustal average, and the soil average are listed in table 2.6. ^{232}Th concentrations range from 2 to 20 ppm in the common rock types, corresponding to activity concentrations of 8 to 80 Bq/kg (0.18 to 22 pCi/g). Like ^{238}U, ^{232}Th has markedly higher concentrations in some parts of the world.

Because of its specific activity and low mobility, except in the low-pH situations mentioned previously, ^{232}Th is normally present in biologic materials only in insignificant amounts. The mean concentration in 25 vegetable samples (Linsalata 1994) was 0.67 ± 0.81 mBq/kg (0.018 ± 0.022 pCi/kg). Thorium was found in the highest concentrations in pulmonary lymph nodes and lungs; this indicates that the principal source of human exposure is inhalation of suspended soil particles (Ibrahim and others 1983; Wrenn and others 1981). Because thorium is removed from bone very slowly, the concentrations of both ^{230}Th (which is found in the ^{238}U decay chain) and ^{232}Th were found to increase with age. Average concentrations of ^{232}Th in major tissues reported by NCRP (1987a) indicated that the highest concentrations (wet weight) were in lung and cortical bone, at 20 and 12 mBq/kg (0.5 and 0.3 pCi/kg), respectively.

The external dose rate due to gamma radiation from the thorium chain is usually somewhat greater than that from the uranium chain and arises primarily from the decay products rather than from ^{232}Th itself. The internal dose from the ^{232}Th chain is due primarily to ^{228}Ra and its decay products (NCRP 1987a), which are discussed in the next section.

Radium-228

Radium-228 is a member of the ^{232}Th chain. Although ^{228}Ra and ^{226}Ra commonly occur in soil and water in about a 1:1 ratio, ^{228}Ra has not been systematically measured in food and water on a scale comparable with that of ^{226}Ra. NCRP (1987a) estimates that the daily intake of ^{228}Ra is about 0.04 Bq (1 pCi), which can be compared to its ^{226}Ra estimate of 0.05 Bq (1.3 pCi). Where elevated concentrations of ^{226}Ra have been noted in drinking water, ^{228}Ra concentrations are often comparable (Hess and others 1985; Gilkeson and others 1984). The geochemistry of ^{228}Ra is essentially identical with that of ^{226}Ra. ^{228}Ra and its decay products are estimated to contribute annual dose equivalents of 300 µSv (30 mrem) to cortical bone, 84 µSv (8.4 mrem) to trabecular bone, 120 µSv (12 mrem) to the bone lining cells, 22 µSv (2.2 mrem) to the red marrow, and 1.5 µSv (0.15 mrem) to soft tissues (NCRP 1987a).

Radon

Radium-226 decays by alpha-particle emission to ^{222}Rn, which has a half-life of 3.82 d. Radium-224, which is a member of the ^{232}Th chain, decays by alpha-particle emission to 55.6 s ^{220}Rn. Radon-219 is a member of the ^{235}U chain and decays most rapidly, having a half-life of 3.96 s. Radon is a noble gas; it occurs as nonpolar, monatomic molecules and is inert for practical purposes. The 3.82-d ^{222}Rn isotope has a greater opportunity than the nuclei of shorter-lived radon isotopes to escape to the atmosphere. The mechanisms by which ^{222}Rn is transported from soil into the atmosphere have been treated extensively by Tanner (1992; 1980; 1964).

When the parent radium decays in rock or soil, the resulting radon atoms recoil and some of them come to rest in geologic fluids, most likely water in the capillary spaces. Some of the radon in soil water enters soil gas, primarily by diffusion, and then becomes more mobile. Radon reaches the atmosphere when soil gas at the surface exchanges with atmospheric gas. A less important mechanism is diffusion from soil gas to atmospheric gas. The concentration of ^{222}Rn in typical soil gas is 4-40 kBq/m^3 (10^2 - 10^3 pCi/L), several orders of magnitude higher than ^{222}Rn concentrations found in the outdoor atmosphere.

Gesell (1983) reviewed the reported data from various parts of the United States and found that the annual average outdoor ^{222}Rn concentration ranged from 0.6 Bq/m^3 (0.016 pCi/L) in Kodiak, AK, to 28 Bq/m^3 (0.75 pCi/L) in Grand Junction, CO, a location with elevated soil radium concentrations. Data from the United States and several other countries indicate that the average concentrations of ^{222}Rn in outdoor air can normally be taken to be 4–19 Bq/m^3 (0.1-0.5 pCi/L). NCRP (1987a) compiled results from 14 studies of outdoor ^{222}Rn concentrations in the United States and found a similar range of 4–18 Bq/m^3 (0.1–0.5 pCi/L), except for Colorado Springs, where the mean for five sites was 44 Bq/m^3 (1.2 pCi/L). Several investigators have determined that the highest concentrations are observed in the early hours and the lowest in the late afternoon, when the concentrations are about one-third the highest morning ones (see for example, UNSCEAR 1982; Gold and others 1964). Over the course of a year, ^{222}Rn concentrations tend to peak in the fall or winter months and have minimums in the spring. This variation is consistent with the pattern of atmospheric turbulence, which tends to be greater in spring.

Because the decay products of ^{222}Rn and ^{220}Rn are electrically charged when formed, they tend to attach themselves to dusts, which are normally present in the atmosphere. If the radioactive decay products of radon are not removed by mechanisms other than radioactive decay, the parents and their various decay products will achieve radioactive equilibrium. The growth of the ^{222}Rn decay products approaches an equilibrium in about 2 h; beyond that, further growth in the activity of the nuclide chain is slowed by the presence of 22.3-y ^{210}Pb, which, in the short term, acts as a nearly stable nuclide. Wilkening (1952) found that the ^{222}Rn decay products tend to distribute themselves on atmospheric dust in a manner that depends on the particle size of the dust, and that the bulk of the activity is contained on particles having diameters less than 0.035 μm.

When air that contains ^{222}Rn or ^{220}Rn in partial or total equilibrium with its decay products is inhaled, the inert gases are largely exhaled immediately. However, some of the dust particles will be deposited in the respiratory system. Additional radon decay products will be deposited with each breath until radioactive equilibrium is reached, at which point the amount of activity deposited per unit time equals the amount eliminated from the lungs by the combination of physiologic clearance and radioactive decay. In the case of ^{222}Rn in equilibrium with its decay products, the total energy dissipation in the lungs derived from the decay products is about 500 times greater than that derived from decay of the ^{222}Rn itself. The dosimetry of radon and its decay products is discussed in Chapter 8.

Indoor Radon In confined spaces, especially those bounded by radon-emitting materials, ^{222}Rn concentrations can be orders of magnitude higher than outdoors. Examples include underground mines (especially uranium mines), caves, and structures, especially one- or two-story homes. One of the surprising developments in recent years has been the finding that in many homes the concentration of ^{222}Rn (and its decay products) is so high as to pose potential risks far greater than those posed by other pollution hazards that have attracted attention. Reviews of indoor ^{222}Rn can be found in Nazaroff and Nero (1988), Nero and others (1990), and Eisenbud and Gesell (1997).

The indoor ^{222}Rn problem exists mainly in homes because the ^{222}Rn originates primarily in the soil, which has its greatest effect on one- or two-story buildings. The building materials themselves are a minor source of ^{222}Rn, compared with soil, except when the materials contain relatively high concentrations of radium and have sufficient permeability and porosity to allow ^{222}Rn to escape. That is true, for example, if gypsum board or another building material has been manufactured as a byproduct of phosphate-fertilizer production (Lettner and Steinhäusler 1988; Paredes and others 1987).

^{222}Rn can enter the indoor atmosphere in a number of ways, including advection and diffusion from soil, diffusion from construction materials, infiltration with outdoor air, emanation from water, and presence in natural gas (UNSCEAR 1988). In EPA's draft report on diffuse NORM waste (EPA 1993b), a diffusion model is used to estimate indoor radon concentrations on the basis of ^{226}Ra concentrations in waste on which a house was built. The model incorporates a one-dimensional version of Fick's law to estimate radon diffusion from soil through concrete of different densities. However, there is evidence that diffusion of ^{222}Rn is a minor pathway compared with the advection of soil gases directly through breaches in the foundation as a result of slight pressure differentials that can result from atmospheric pressure changes, temperature differentials, or wind velocity. For example, the UN Scientific Committee on the Effects of Atomic Radiation (UNSCEAR 1988) has shown that advection typically accounts for 75% of the radon that enters a reference house, whereas diffusion accounts for only 3%. Steinhäusler (1975) has shown that meteorologic factors in particular can influence indoor concentrations of ^{222}Rn and its decay products. The approximate contributions of various sources to the indoor ^{222}Rn concentrations of single-family dwellings and apartments are given in table 2.7 (Nero 1988; Nero and others 1986).

Several efforts to estimate the US national distribution of indoor ^{222}Rn have been made (Marcinowski and others 1994; White and others 1992; Cohen 1991; Cohen 1989; Alter and Oswald 1987; Cohen 1986; Nero and others 1986), but the most current representative US survey of indoor ^{222}Rn is the National Residential Radon Survey (Marcinowski and others 1994; EPA 1992b). From this survey, the average national ^{222}Rn concentration was found to be 46 Bq/m^3 (1.25 pCi/L) and the median, 25 Bq/m^3 (0.67 pCi/L). The average and median ^{222}Rn concentrations in each of the 10 EPA regions are shown in figure 2.1. Regionally, the Midwest and Intermountain West have the highest indoor ^{222}Rn concentrations, averaging about twice the national average, whereas the Northwest has the lowest. Figure 2.2 shows the distribution of ^{222}Rn

Table 2.7 Approximate contributions of various sources to indoor ^{222}Rn[a]

Source	Single-family homes		Apartments (high-rise)	
	pCi/L	Bq/m^3	pCi/L	Bq/m^3
Soil (based on flux measurements)	1.5	55	>0	>0
Water (public supplies)[b]	0.01	0.4	0.01	0.37
Building materials	0.05	2	0.1	3.7
Outdoor air	0.25	10	0.25	9.25
Observed indoor concentrations	1.5	55	0.3 ?	12 ?

[a] Adapted from Nero (1988), Nero and others (1986).
[b] Applies to residences served by public supplies. Contributions can be much larger in some cases.

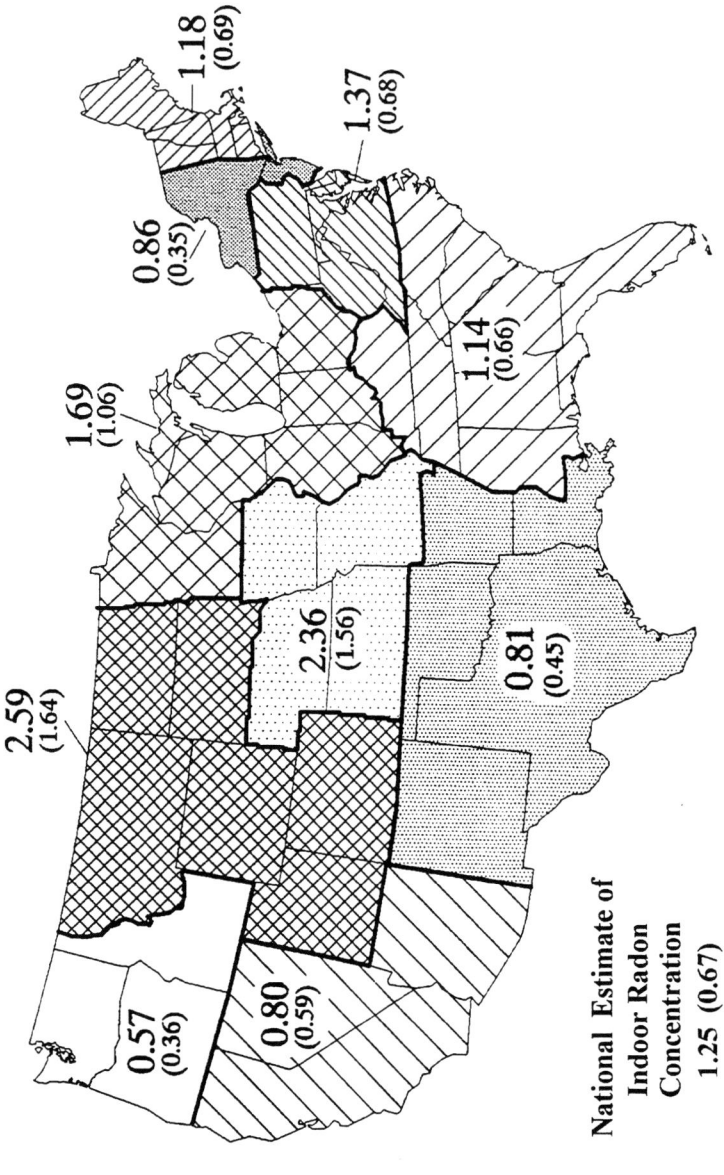

Figure 2.1. Mean and median (in parentheses) ^{222}Rn concentrations in all living areas by EPA region (in pCi/L; 1 pCi/L equals 37 Bq/m^3). (From EPA 1992b).

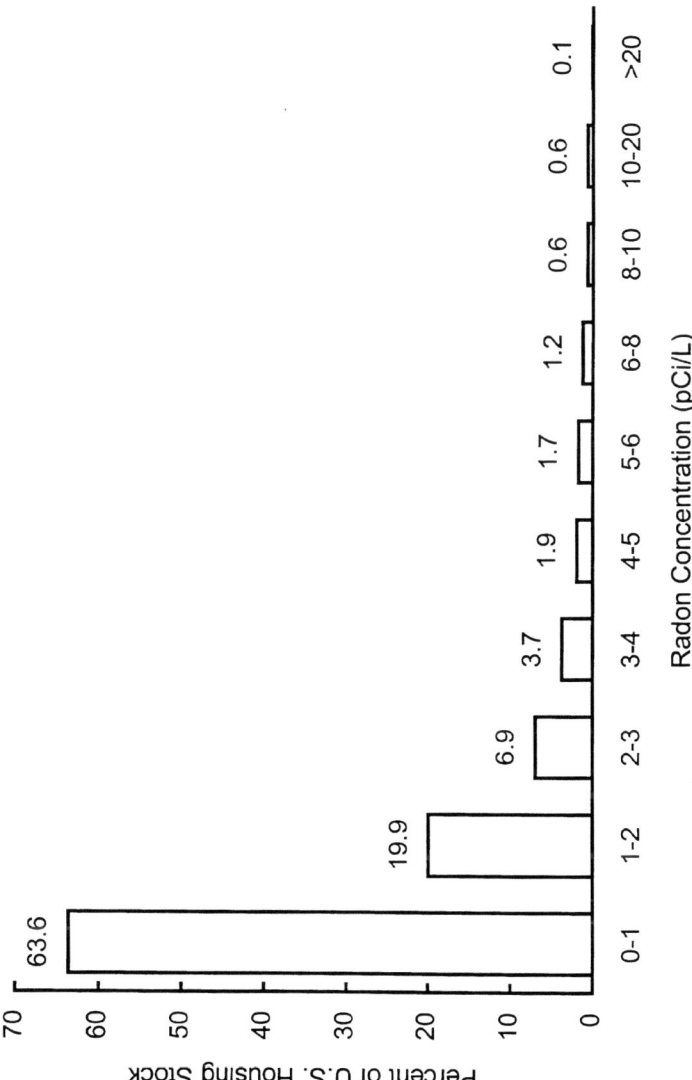

Figure 2.2. National distribution of ^{222}Rn concentrations (pCi/L) in US housing stock averaged over all living areas (1 pCi/L = 37 Bq/m^3). (From EPA 1992b).

concentrations for the entire country. About 6% of the residences surveyed had ^{222}Rn concentrations over 150 Bq/m^3 (4 pCi/L). Applied nationally, that implies that 5.8 million residences have ^{222}Rn concentrations exceeding 150 Bq/m^3 (4 pCi/L). The results from the Nero and others (1986) review, the Cohen (1986) study, and the carefully designed National Residential Radon Survey (EPA 1992b) are similar, suggesting that the distribution of indoor ^{222}Rn in the United States is reasonably well characterized.

World indoor ^{222}Rn concentrations do not necessarily follow the pattern seen in the United States. They are often higher in Scandinavian countries, such as Denmark (NIRH 1987), where the average of the summer and winter ^{222}Rn concentrations is 93 Bq/m^3 (2.5 pCi/L). A very low result was seen in Australia in a nationwide ^{222}Rn survey of homes (Langroo and others 1991).

On the basis of limited measurements in a few buildings (Turk and others 1986; Cohen and others 1984) and reasoning that multistory buildings with forced ventilation would be less likely to reach high ^{222}Rn concentrations (Nero 1988), ^{222}Rn concentrations in commercial and industrial structures were generally believed to be much lower than those in residences. However, they might warrant reconsideration. High ^{222}Rn has been found in underground workplaces in Germany (Schmitz and Fritsche 1992). Scott (1992) identified 86 buildings at seven DOE sites that might exceed the EPA action level of 150 Bq/m^3 (4 pCi/L) for residences. That amounts to 2.8% of the 3,100 structures surveyed, about half the percentage of US residences estimated to exceed 150 Bq/m^3 (4 pCi/L).

Natural underground caves have limited ventilation and are bounded by rock and soil capable of emanating radon into the air. Radon is also carried into caves by water. ^{222}Rn concentrations are typically much higher in caves than outdoors (Wilkening and Watkins 1976). In a study of caves operated by the US National Park Service, Yarborough (1980) identified numerous locations in several caves with radon greater than 7.5 kBq/m^3 (200 pCi/L).

Radon in Groundwater ^{222}Rn dissolved in potable water is another source of human exposure, mainly because the ^{222}Rn is released from solution at the tap and enters the home atmosphere (Nazaroff and others 1987; Watson and Mitsch 1987; Cross and others 1985; Prichard and Gesell 1983; Gesell and Prichard 1975). Water supplies ordinarily make only a small contribution to the indoor ^{222}Rn concentration but can be the predominant source in areas where the ^{222}Rn content of groundwater is unusually high. Studies in Maine and Colorado have shown ^{222}Rn in water to be an important contributor in some dwellings (Lawrence and others 1992; Hess and others 1981).

In the 1992 Lawrence and others study, performed in Colorado, estimates of the concentration of indoor radon attributed to radon in the domestic water supply depended on assumptions regarding the fraction of radon emanating from the water and on dwelling ventilation rates. The averages of the concentrations of indoor radon attributed to radon in the domestic water supply for the 28 houses studied were 20 Bq/m^3 (0.54 pCi/L) with assumptions that minimized radon attributable to water and 48 Bq/m^3 (1.3 pCi/L) with assumptions that maximized radon attributable to water. The highest estimated value of indoor radon attributed to radon in the domestic water supply for a single dwelling was 310 Bq/m^3 (8.4 pCi/L). The proportion of total indoor radon concentration attributable to radon in water was estimated to range up to 77%. Continuous measurements in a single house demonstrated a strong correlation between water use and indoor radon concentration.

Lead-210 and Polonium-210

Lead-210 is a 22.3-y beta-particle emitter separated from its antecedent ^{222}Rn by six short-lived alpha-particle and beta-particle emitters (see table 2.3). The longest-lived radionuclide between ^{222}Rn and ^{210}Pb is ^{214}Pb, which has a half-life of only 26.8 min. ^{210}Pb decays to 138.4-d ^{210}Po via the intermediate ^{210}Bi, which has a 5 d half-life (see table 2.3). Thus, after the decay of 3.82-d ^{222}Rn in the atmosphere, ^{210}Pb is produced rapidly, but its long half-life allows little to decay in the atmosphere before it precipitates to the earth's surface, mainly in rain or snow.

The ^{210}Pb content of the atmosphere has been found to vary from 0.2 to 1.5 mBq/m^3 (5 x 10^{-3} to 40 x 10^{-3} pCi/m^3), with the lowest values at such island stations as San Juan, PR, and Honolulu, HI, and the highest values in the interior of the United States (NCRP 1987a). The mean residence time of dust suspended in the troposphere is about 15 d, so there is little time for ^{210}Po to be formed in suspended dust, and the concentration of ^{210}Po near ground level is smaller than that of ^{210}Pb. For purposes of estimating dose in the United States, NCRP (1987a) has adopted nominal ground-level concentrations for ^{210}Pb and ^{210}Po of 0.7 and 0.07 mBq/m^3 (20 x 10^{-3} and 2 x 10^{-3} pCi/m^3), respectively. On the basis of few measurements, Fisenne (1993) estimated that ^{210}Pb concentrations indoors are about one-fourth those outdoors. NCRP (1987a) has estimated that the mean dietary intake of ^{210}Pb is about 0.05 Bq/d (1.4 pCi/d) and that the ^{210}Po content of the standard diet is an average of 1.3 times that of ^{210}Pb. Food and water ingestion is a more important contributor to blood ^{210}Pb than inhalation. In the United States, ^{210}Pb and its decay products are estimated to contribute an annual equivalent dose of 1,400 µSv (140 mrem) to cortical and trabecular bone, 700 µSv (70 mrem) to the bone lining cells, and 140 µSv (14 mrem) to the red marrow and soft tissues (NCRP 1987a).

In two population groups, ^{210}Pb and ^{210}Po concentrations are apt to be higher than average: cigarette-smokers and people who eat substantial quantities of caribou from northern lands. ^{210}Pb and ^{210}Po are believed to enter tobacco by being deposited on tobacco leaves from the atmosphere (Martell 1974). When the tobacco is smoked, the ^{210}Pb and ^{210}Po are volatilized and inhaled, and this results in blood concentrations of these nuclides about one-third higher than in nonsmokers. Caribou feed on lichens, which absorb trace elements in the atmosphere, including ^{210}Pb and ^{210}Po. The tissue content of these nuclides in Lapps in northern Finland, who subsist on caribou, was about 12 times that of residents of southern Finland, where normal Scandinavian dietary regimes exist (Persson 1972; Kauranen and Miettinen 1969).

Potassium-40

Of the three naturally occurring potassium isotopes, only ^{40}K is unstable, having a half-life of 1.3×10^9 y. It decays by beta-particle emission to calcium-40 (89%) and by electron capture to argon-40 (11%) and produces 1.46-MeV gamma rays after electron-capture decay. Potassium-40 is present at 0.0117% by mass in natural potassium, thereby imparting a specific activity of about 30 kBq/kg (800 pCi/g) of potassium. Representative values of the total potassium content of rocks, as summarized in table 2.6, indicate a wide range of values, from 0.3% to 4.5% for various rock types. That corresponds to an activity concentration range of 90 to 1,400 Bq/kg (2.5 to 37 pCi/g). Some basalts and sands are low in potassium, whereas granites and other basalts are high. It has been estimated that about 110 TBq (3,000 Ci) of ^{40}K is added annually to the soils of the United States in the form of fertilizer (Guimond 1978). Seawater contains ^{40}K at about 11 kBq/m^3 (300 pCi/L). Because of its relative abundance and its energetic beta-particle emission (1.3 MeV), ^{40}K is the predominant radioactive component in common foods and human tissues. It is important to recognize that the potassium content of the body is under homeostatic control and is little influenced by environmental variations. The dose from ^{40}K in the body is therefore reasonably constant. A person who weighs 70 kg contains about 140 g of potassium, most of which is in muscle. From the specific activity of potassium, it follows that the ^{40}K content of the human body is around 4 kBq (0.1 µCi). NCRP (1987a) has estimated that this radionuclide delivers an annual dose of 0.18 mSv (18 mrem) to the soft tissues and 0.14 mSv (14 mrem) to bone. However, Paschoa and others (1992) have questioned the conventional dosimetry of ^{40}K and other nuclides that decay by electron capture because the intracellular dose from Auger electrons, which have energies of a few thousand electron volts, has not been considered.

Rubidium-87

The primordial beta-emitting radionuclide ^{87}Rb, with a half-life of 4.75 x 10^{10} y, is present in the environment and in human tissues at low concentrations. Estimates of the average annual effective dose equivalent from ^{87}Rb are 3-6 μSv (0.3-0.6 mrem) (UNSCEAR 1988; NCRP 1987a).

Induced Radionuclides

Some radionuclides that exist on the surface of the earth and in the atmosphere have been produced by the interaction of cosmic rays with atmospheric nuclei. The two most important of these induced radionuclides, tritium (^{3}H) and ^{14}C, are only minor dose contributors relative to the primordial radionuclides discussed in previous sections. Some of the properties of these radionuclides and the extent to which they have been reported in various media are listed in table 2.1. It is estimated that the annual dose from ^{14}C is 30 μSv (3 mrem) to the skeletal tissues of the body and 10 μSv (1 mrem) to the soft tissues. The annual average dose from ^{3}H of natural origin is estimated (NCRP 1987a) at 0.01 μSv (1 μrem).

NATURAL SOURCES OF EXTERNAL IONIZING RADIATION

The dose received from external sources of ionizing radiation originates in cosmic rays and photon-emitting radionuclides in the earth's crust (terrestrial sources).

Terrestrial Sources of External Radiation

The major terrestrial sources of gamma radiation are ^{40}K and nuclides of the ^{238}U and ^{232}Th chains. The relationship between soil concentration of radionuclides and dose was developed originally by Hultqvist (1956) and by Beck (1980; 1975). Tables relating external dose rates to the concentrations and distributions of many radionuclides have been published as *Federal Guidance Report No. 12* by Eckerman and Ryman (1993). The annual effective dose equivalent 1 m above soil that has uniformly distributed typical concentrations of each of the three major sources of terrestrial radiation is shown in table 2.8. For a hypothetical, unshielded individual residing full-time on a potassium chloride salt-flat, the maximum annual external dose would be about 5.4 mSv (540 mrem).

Extensive measurements of the natural gamma-radiation background in a number of cities throughout the United States show (figure 2.3) that natural radiation exposure rates from terrestrial sources in the United States vary from less than 1 to about 20 μR/h (Beck 1966). The temporal variation is illustrated by a series of measurements performed by the Environmental Measurements

Table 2.8 Estimated Effective Dose Equivalents 1 Meter above Soil Containing Uniformly Distributed Typical Concentrations of Important Natural Radionuclides.

Nuclide	Typical Soil Concentration[a], Bq/kg	Dose Coefficient[b], µSv/y per Bq/kg	Annual Effective Dose Equivalent, µSv
^{40}K	400	0.28	112
^{238}U + decay products	22	3.05	67
^{232}Th + decay products	37	4.37	162

[a]From NCRP (1987a).
[b]Calculated by using data from *Federal Guidance Report No. 12* (Eckerman and Ryman 1993)

GUIDELINES FOR EXPOSURE TO TENORM 51

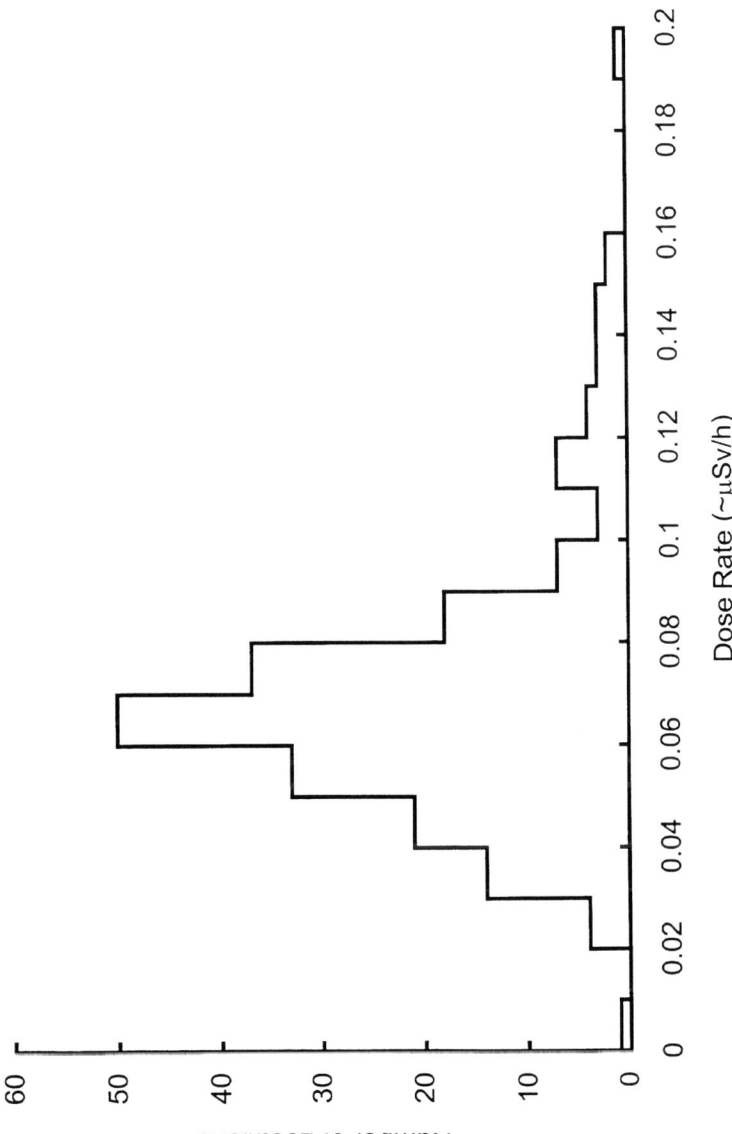

Figure 2.3. Frequency distribution of gamma dose rate from natural emitters at 210 locations in United States. (From Beck 1966).

Laboratory at its rural background monitoring station in Chester, NJ (Klemic 1996). Figure 2.4 shows short-term variations and the effects on dose rate of diurnal variations in radon concentration, soil moisture, and rainout of radon decay products. Diurnal variations in radon concentration are caused by diurnal changes in atmospheric stability. Rainout of radon decay products briefly increases the dose rate, whereas accumulated soil moisture decreases it as a result of attenuation of the gamma-ray flux. Figure 2.5 shows long-term variations, which are influenced mostly by the attenuating effects of soil moisture and snow cover.

In addition to calculations and direct ground-level measurements of external dose, measurements can be made with sensitive gamma-ray detectors in aircraft (IAEA 1991). Many such surveys have been made, either to explore for uranium or to provide information about the radiation in the vicinity of proposed nuclear facilities. The data were analyzed by Oakley (1972), who estimated the population dose distribution in the United States. The data are grouped by geographic region: (1) the Atlantic and Gulf coastal plain, for which the mean annual absorbed dose is 0.23 mGy (23 mrad); (2) a portion of the eastern slope of the Rocky Mountains, where the annual absorbed dose averages 0.9 mGy (90 mrad); and (3) the remainder of the United States, where the average annual absorbed dose is 0.46 mGy (46 mrad).

Cosmic Radiation

The primary radiation that originates in outer space and impinges isotropically on the top of the earth's atmosphere consists of 87% protons, 11% alpha particles, about 1% nuclei of elements of atomic number 4-26, and about 1% electrons of very high energy. An outstanding characteristic of the cosmic radiation is that it is highly penetrating, with a mean energy of about 10^{10} eV and maximum energy of as much as 10^{20} eV. The primary radiation predominates in the stratosphere above an altitude of about 25 km (NCRP 1987a).

Most cosmic radiation originates outside the solar system. However, the solar component is important outside the atmosphere after flares associated with sunspot activity that follows an 11-y cycle.

The interactions of the primary particles with atmospheric nuclei produce electrons, gamma rays, neutrons, pions and muons. At sea level, muons account for about 80% of the cosmic-radiation charged-particle flux, and electrons account for about 20%. The neutron flux is comparable with the electron flux.

GUIDELINES FOR EXPOSURE TO TENORM 53

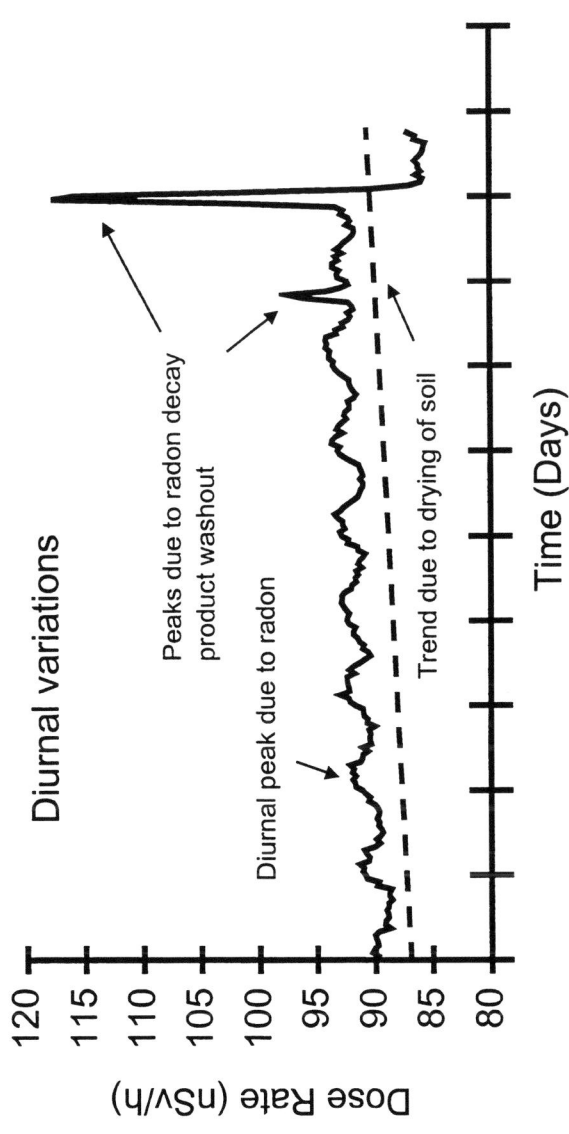

Figure 2.4. Short-term variations in external dose rate at rural site not influenced by industrial sources. Measurements made with pressurized ionization chamber. Arrow identifies one of nightly maximums attributed to increase in ^{222}Rn concentration. Dashed line shows gradual increase attributed to soil drying. Large peak is typical of washout and subsequent decay of short-lived radon decay products during rainfall; lower baseline after peak is caused by attenuating effect of increase in soil moisture. (From Klemic 1996).

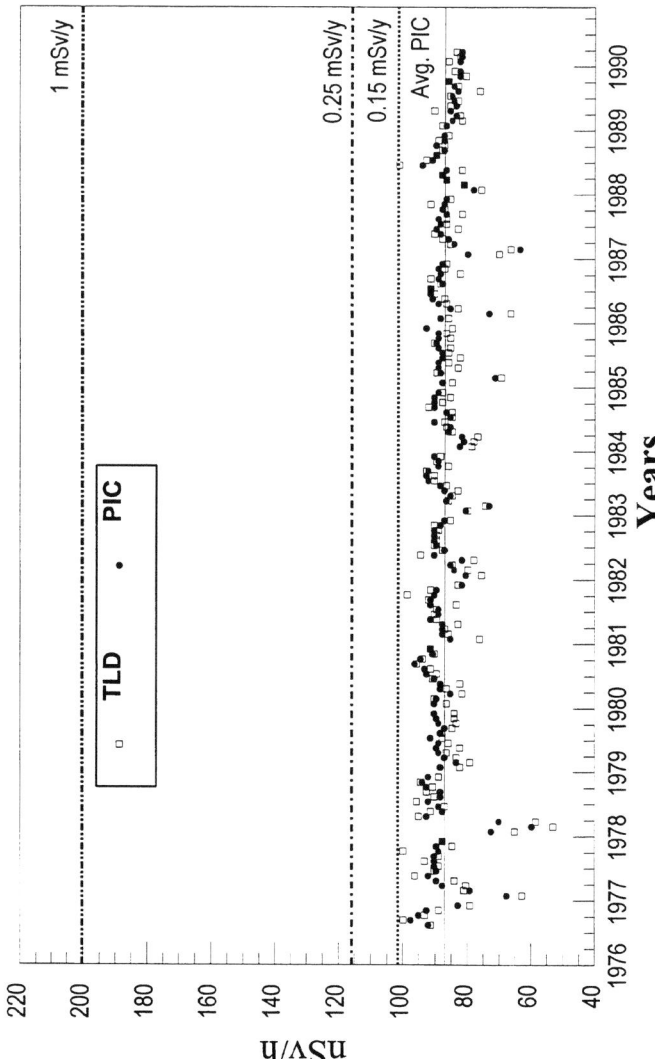

Figure 2.5. Long-term variations in external dose rate at rural site not influenced by industrial sources. Measurements made with pressurized ionization chamber and thermoluminescent dosimeters. Arrow identifies one of the nightly maximums attributed to increase in ^{222}Rn concentration. (From Klemic 1996).

GUIDELINES FOR EXPOSURE TO TENORM 55

The dose from cosmic radiation is markedly affected by elevation. The annual cosmic-ray dose equivalent is about 0.29 mSv (29 mrem) at sea level. For the first few kilometers above the earth's surface, the cosmic-ray dose rate doubles for each 2,000-m increase in altitude (figure 2.6).

With the development of high-altitude aircraft and manned space flight, the dose from primary cosmic radiation attracted interest (O'Brien and McLaughlin 1972; Curtis and others 1966), which continues to the present (NCRP 1995; Reitz and others 1993; NCRP 1989b). A transcontinental flight has been estimated to result in a dose of about 0.025 mSv (2.5 mrem), or 0.05 mSv (5 mrem) per round trip (NCRP 1987a). Air crews who work an exceptionally heavy schedule (1,100 h/y) can receive annual doses of 0.3-9 mSv (30-900 mrem), depending on the routes flown (O'Brien and others 1992). Once or twice during the 11-y cycle, a giant solar event can deliver dose equivalents at very high altitudes (15-25 km) of 10-100 mSv/h (1-10 rem/h), with a peak as high as 500 mSv (5 rem) during the first hour (Upton and others 1966). During a well-documented solar flare in February 1956, dose rates in excess of 1 mSv/h (100 mrem/h) existed briefly at altitudes as low as 10,000 m (Schaefer 1971).

SUMMARY OF HUMAN EXPOSURES TO NATURAL IONIZING RADIATION

The annual effective dose equivalent received by persons living in areas of normal background radiation is estimated at 2.4 mSv (240 mrem) for the world population (UNSCEAR 1988). The annual external effective dose equivalent is estimated at 0.36 mSv (36 mrem) from cosmic sources and 0.41 mSv (41 mrem) from terrestrial radiation. ^{222}Rn and its short-lived decay products contribute about 40% of the total effective dose equivalent. The natural sources of dose are shown in more detail in table 2.9.

A somewhat larger total annual dose of 3 mSv (300 mrem) is estimated for residents of the United States and is shown in detail in table 2.10 (NCRP 1987a). The US estimates are 0.27 mSv (27 mrem) for cosmic sources and 0.28 mSv (28 mrem) from terrestrial radiation. The major difference between the two estimates, however, is the average effective dose equivalent due to ^{222}Rn, which is 55% of the total in the US estimate but 40% of the total for the UNSCEAR estimate. That is a difficult quantity to estimate, because world average ^{222}Rn concentrations are not well known and several models are used to convert ^{222}Rn exposure to lung dose (chapter 8).

The population distribution of external dose in the United States from terrestrial and cosmic sources combined is shown in figure 2.7 and is seen to range over a factor of about 4. The variation in radon exposure would be

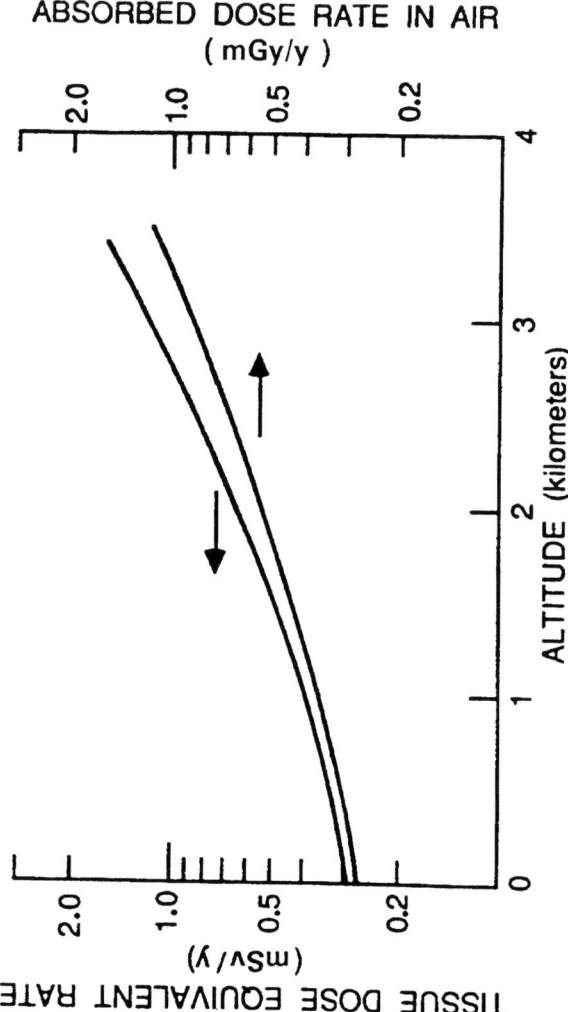

Figure 2.6. Variation in cosmic-ray dose with elevation. Charged-particle absorbed dose rate in air or tissue shown in lower curve; total dose equivalent rate (charged particles plus neutrons) shown in upper curve at depth of 5 cm in 30-cm-thick slab of tissue. Dose rates can be converted to mSv/y or mGy/y by dividing by 100. (From NCRP 1987a).

Table 2.9 Estimated Effective Dose Equivalents From Natural Sources in Normal Regions[a]

Source	Annual Effective Dose Equivalent					
	mrem			mSv		
	External	Internal	Total	External	Internal	Total
Cosmic, including neutrons	36	--	36	0.36	--	0.36
Cosmogenic nuclides	--	1.5	1.5	--	0.015	0.015
Primordial nuclides						
^{40}K	15	18	33	0.15	0.18	0.33
^{87}Rb	--	0.6	0.6	--	0.006	0.006
^{238}U chain						
^{238}U $\Rightarrow$ ^{234}U	--	0.5	0.5	--	0.005	0.005
^{230}Th	--	0.7	0.7	--	0.007	0.007
^{226}Ra	10	0.7	10.7	0.1	0.007	0.107
^{222}Rn $\Rightarrow$ ^{214}Pb	--	110	110	--	1.1	1.1
^{210}Pb $\Rightarrow$ ^{210}Po	--	12	12	--	0.12	0.12
^{232}Th chain						
^{232}Th	--	0.3	0.3	--	0.003	0.003
^{228}Ra $\Rightarrow$ ^{224}Ra	16	1.3	17.3	0.16	0.013	0.173
^{220}Rn $\Rightarrow$ ^{208}Pb	--	16	16	--	0.16	0.16
Total (round)	80	160	240	0.8	1.6	2.4

[a]Adapted from UNSCEAR (1988).

Table 2.10 Estimated Effective Dose Equivalent Received by Member of US Population[a]

Source	Average Annual Effective Dose Equivalent	
	mrem	mSv
Inhaled radon and decay products	200	2
Cosmogenic radionuclides (such as ^{14}C)	1	0.01
Other internally deposited radionuclides (such as ^{40}K and ^{210}Po)	40	0.40
Terrestrial radiation	28	0.28
Cosmic radiation	27	0.27
Total (round)	300	3

[a]Adapted from NCRP (1987a).

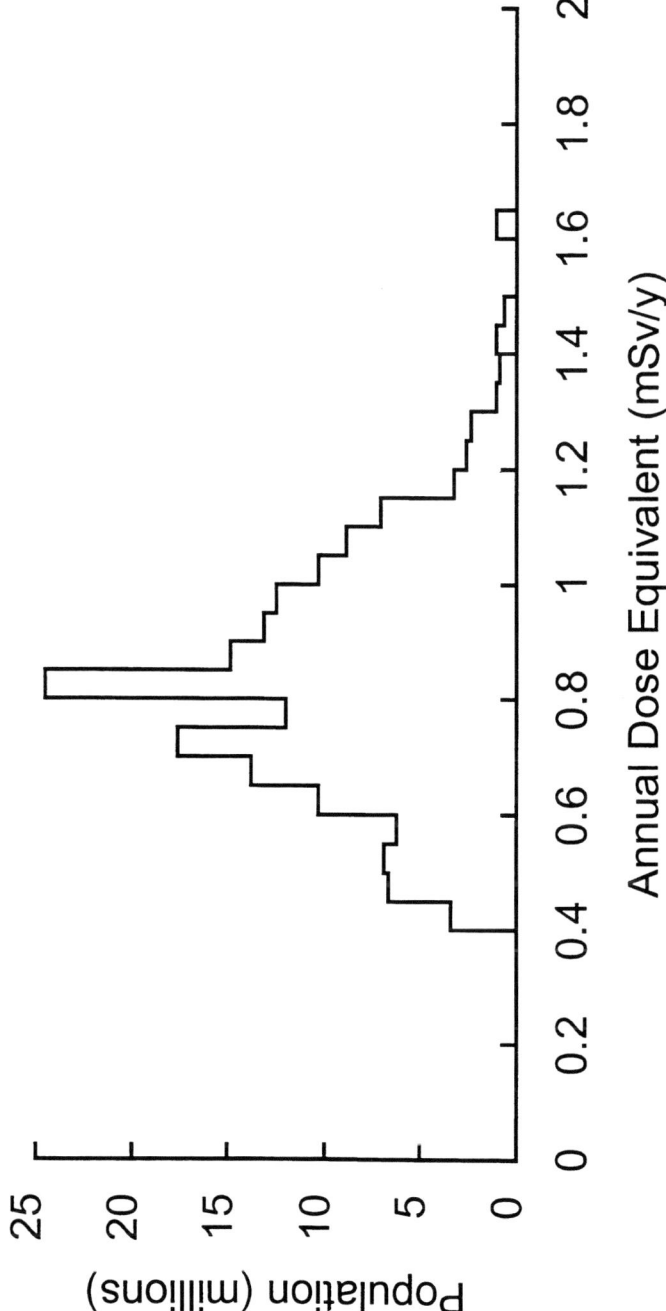

Figure 2.7. Distribution of total external dose rate (terrestrial and cosmic) by population for United States. (From Oakley 1972).

expected to be nearly proportional to the distribution of indoor radon levels in figure 2.2, which implies a range of factor of more than 20.

CONCLUSIONS

The main conclusions drawn from the foregoing review are as follows:

- All natural media—earth, air, water, and biota, including humans—are radioactive to some degree, and the concentrations of radionuclides in these media are highly variable, both between and within media.

- Humans receive radiation exposure from natural sources outside and inside the body, averaging about 1 mSv (100 mrem) per year in the United States.

- Humans receive radiation exposure from radon averaging about 2 mSv (200 mrem) per year in the United States.

- Doses received by humans from sources of natural radiation in the environment are quite variable, with a range of a factor of about 4 for external sources except radon and about 20 for radon.

As a practical matter, the implications of existing levels and the variability of natural radionuclides and doses received by humans should receive careful consideration as efforts to regulate TENORM are contemplated.

3
Major Sources of Technologically Enhanced Naturally-Occurring Radioactive Materials

TENORM spans a wide spectrum of raw materials and products destined for use, recycling, or disposal. This chapter summarizes TENORM concerns associated with various industrial activities and notes unique characteristics of possible importance in dose assessment. Because of the diversity of sites, materials, and processes (table 3.1), it is difficult to summarize radionuclide concentrations and waste volumes here. For that kind of information, the reader is referred to an Environmental Protection Agency (EPA) review (EPA 1993b).

URANIUM MINING

Uranium production from surface mining operations generates large volumes of overburden with either ambient or elevated, but below-ore-grade, concentrations of uranium and its decay products. Smaller amounts of waste rock are produced by underground uranium mines. The ratio of overburden to ore has increased as less-accessible and lower-grade ores have been exploited. In the 1950s, the ratio was about 10:1; by the 1980s, it had increased to about 60:1. Most of the mines in question are in the western states: Arizona, Colorado, New Mexico, South Dakota, Texas, Utah, and Wyoming. A 1989 survey showed the average radium-226 concentration in uranium-mine overburden to be about 0.9 kBq/kg (25 pCi/g).

Those mining wastes are distinct from uranium mill tailings (UMT), which are the ore residues discharged to a waste pond after extraction of the uranium, typically by sulfuric acid leaching. Although 90-95% of the uranium in the ore is extracted, most of the uranium-decay-product activity remains with

Table 3.1 Domestic Industrial Activities That Generate TENORM[a]

Industrial Activity	TENORM Form	Major Generator Locations
Uranium mining	Overburden	Arizona, Colorado, New Mexico, South Dakota, Utah, Wyoming, Texas
Phosphate-fertilizer and elemental-phosphorus production	Ferrophosphorus Phosphogypsum Scale Slag	Florida, Idaho, other states in West and Southeast
Coal combustion	Fly ash Bottom ash and slag	Widespread
Oil and natural-gas production and processing	Scale Sludge Contaminated water and production and processing equipment	All petroleum and natural gas states that have production and processing-facilities
Municipal water treatment	Sludge Radium-selective resins	North central (such as Illinois), Coastal Plain (such as North Carolina), other states

[a]Modified from (DOE 1996).

Table 3.1 (continued)

Industrial Activity	TENORM Form	Major Generator Locations
Metal mining and processing	Slag, leachate, scale, and tailings from:	
	Rare earth metals	California, Florida, North Carolina
	Special-application metals (zirconium, hafnium, titanium, and tin)	Ohio, Delaware, Florida, Oregon
	Large-volume metal-processing industries (copper, iron, and so on)	Ohio, Pennsylvania, Indiana, Illinois, Michigan, some western states
Geothermal energy production	Scale Brine-pond deposits	California
Others:		
Pulp and paper production	Scale	
Metal casting	Zircon sand ceramic shells	
Former radium-processing and manufacturing facilities	Contaminated soils	Colorado, Pennsylvania, New Jersey, Illinois

the UMT. UMT are regulated under EPA's standards for uranium and thorium mill tailings (40 CFR Part 192) and are therefore not a focus of this report. Some of the TENORM considered in this report has had processing and disposal histories similar to those of UMT, and UMT have been the focus of more research than most other TENORM. However, such UMT properties as radon emanation coefficients and leachability of radionuclides should not be generalized to the entire TENORM spectrum of materials without due consideration of material similarities and differences.

PHOSPHATE-FERTILIZER AND ELEMENTAL-PHOSPHORUS PRODUCTION

Up to about 0.02% uranium can substitute for positions typically occupied by atoms of calcium in the structure of the mineral carbonate fluorapatite (Durrance 1986). This mineral commonly occurs in phosphate rock, the ore for the production of phosphoric acid and elemental phosphorus. Commercial extraction of uranium has occurred at several phosphoric acid plants in Florida and Louisiana (DOE 1996). The transfer of uranium-series radionuclides to both the waste materials and the fertilizer product makes each of these a diffuse source of TENORM. Phosphate operations in Florida amount to about 80% of domestic production; other major mining and processing plants are in Idaho, Louisiana, Mississippi, North Carolina, and Wyoming.

Two distinct manufacturing processes, a wet process and a thermal process, are involved. The wet process treats the ore with sulfuric acid to yield phosphoric acid and hydrated calcium sulfate (gypsum). This "phosphogypsum" (PG) waste contains trace amounts of $^{226}RaSO_4$ coprecipitated with the $CaSO_4 \cdot n$ H_2O. About 80% of the ^{226}Ra in the ore follows the PG, which results in an average ^{226}Ra concentration of about 1.1 kBq/kg (30 pCi/g). The volumes of the waste are large: the process yields about 5 tons of PG for every ton of phosphoric acid produced. The PG waste has been disposed of both on land and in rivers. Land disposal generally results in large piles (called "gyp stacks"). Hull and Burnett (1996) found process waters contained in these stacks to have high ionic strength, low pH (1.4-2.5), and low concentrations of ^{226}Ra (0.08-0.30 Bq/L), because of high sulfate concentrations, but high concentrations of uranium and lead-210 (up to 65 Bq/L). Studies on the Mississippi River and the Rhine estuary have shown that where PG is discharged to such fresh or brackish water, the gypsum dissolves rapidly, releasing the radium, which remains in solution or sorbs onto suspended sediment (Pennders and others 1992; Kraemer and Curwick 1991). Lead-210 and polonium-210 in the PG occur as an insoluble residue that settles in the estuarine discharge zone, whereas the gypsum dissolves. The mineralogic occurrence of the ^{210}Pb and ^{210}Po in the PG is probably similar to that observed by Landa and others (1994) in acid-leach UMT effluent—lead sulfate crystallites in a gypsum matrix. Sediment enrichment in uranium has also been seen in a tidal marsh in Spain where phosphate-fertilizer manufacturing plants discharge liquid and solid wastes to a river (Martinez-Aguirre and others 1996).

About 85% of the uranium partitions to the phosphoric acid, which is further processed to produce a variety of fertilizers with uranium-238 concentrations of about 740-2200 Bq/kg (20-60 pCi/g). Gypsum is often applied to soils as a fertilizer source of calcium and sulfur. PG is an inexpensive and readily available byproduct source of gypsum for agricultural uses. To limit radiation exposure (principally by direct gamma radiation and indoor-radon inhalation exposure), EPA (1992d) issued a ruling that bans the agricultural use of PG that contains ^{226}Ra at over 370 Bq/kg (10 pCi/g). Pipe scales that contain ^{226}Ra at up to 3.7 x 10^3 kBq/kg (1 x 10^5 pCi/g) are a low-volume, discrete TENORM waste at wet-process plants.

The thermal process involves heating the phosphate rock to about 1300 °C to yield an elemental phosphorous product and a calcium silicate vitreous slag waste containing ^{226}Ra at about 0.7-2 kBq/kg (20-50 pCi/g). The low coefficient of radon emanation from this glassy material (the fraction of the radon formed in a radium-bearing solid that escapes to the atmosphere) limits the radon-exposure pathway. Another atmospheric pathway involves the volatilization of ^{210}Pb and ^{210}Po associated with the heating of the ore (EPA 1989d); such releases are regulated under the Clean Air Act (see chapter 7). At a thermal plant in the Netherlands, where such stack off-gases are vented through a wet scrubber for emission control, the water from the system is discharged to an estuary. In sharp contrast with the case of the wet-process PG effluent noted above, about 30% of the ^{210}Pb and 10% of the ^{210}Po are dissolved in the thermal-plant effluent; through dilution with seawater by a factor of about 200, the ^{210}Po figure increases to 100% (Pennders and others 1992). The bioavailability of a soluble radionuclide can be expected to be much higher than that of its insoluble form, so the importance of understanding TENORM processing and geochemical forms of radionuclides when doing dose assessments is clear.

The application of phosphate fertilizers to soils may increase their uranium and radium content. Over 50-80 y of application, the concentrations of ^{238}U and ^{226}Ra in the plow layer could be increased from a few percent to several times background (NCRP 1987b; Pfister and others 1976).

RESIDUES OF COAL COMBUSTION

The reducing conditions under which coals form are conducive to the accumulation of uranium. Typical mucks, peats, lignites, and coals contain uranium at about 0.05-3 ppm. Thorium is strongly adsorbed by peats, and the typical coal contains thorium at 1-10 ppm (Boyle 1982). On combustion of coals, most of the uranium, thorium, and decay products remain with the ash. For its evaluation of coal ash, EPA (1993b) considered composited fly ash plus bottom ash with a literature-derived mean ^{226}Ra concentration of about 0.14 kBq/kg (3.7 pCi/g). Although most coals have decay products in secular equilibrium with the parent, a young, postglacial peat deposit in northeastern Washington state with about 0.1% uranium has less than 10% ingrowth of the possible decay-product activity. This deposit has been exploited as a uranium source, and the lack of decay-product activity rendered the UMT here more benign than those at a typical uranium mill—a factor that was considered in the licensing decision (Stohr and Erickson 1984). Disequilibrium can also be seen in combustion products. Data on fly ash presented by Baxter (1996) suggest strong enrichment (with respect to other uranium-series nuclides) in ^{210}Pb and ^{210}Po, presumably because of volatilization and subsequent condensation on the fly-ash particles. The bottom ash is assumed to show depletion in these radionuclides. Such differential behavior and the resulting concentration differences should be considered in dose assessments.

Radon emanation from ash is a possible exposure pathway from both ash disposal piles and use of fly ash as a concrete aggregate. The low coefficients of radon emanation from glassy materials, such as coal ash and phosphate slag, mitigate exposures.

OIL AND NATURAL-GAS PRODUCTION AND PROCESSING

Oil and natural-gas reservoirs commonly contain large quantities of saline water. These brines come to the surface during pumping operations and require disposal after separation of the water from the oil and gas. The disposal of oilfield brines in a manner that does not result in the salinization of soil and water has been a concern since the early days of the petroleum industry. The radiation hazard was recognized later; the brines tend to be low in uranium (because of reducing conditions in the petroleum reservoir) and low in thorium, but they can contain elevated concentrations of ^{226}Ra and ^{228}Ra (Perel'man 1977). In the United States, more than 90% of such brine is disposed by injection underground, sometimes in enhanced oil-recovery wells, and at other times solely with waste disposal as a goal. The remainder is disposed by surface discharge to earthen evaporation or seepage pits or to wetlands, streams, and so on (Smith 1992). Some brines have been applied to dirt roads for dust control (Rittiger and Yusko 1996). At offshore wells on the Outer Continental Shelf of the Gulf of Mexico, overboard disposal of produced well solids (formation sand, TENORM scale, and so on) is banned, but overboard disposal of production water (treated to remove such solids) is allowed (Minerals Management Service 1996a).

At some facilities in Texas, radium has been removed from brines by treatment with activated charcoal prepared from walnut hulls. The spent charcoal is thus rendered a solid TENORM waste (Ruth McBurney, Texas Department of Health, personal communication, 1993). In Pennsylvania, a brine-treatment facility using pH adjustment and flocculation techniques to remove metals yielded a sludge that was dewatered and sent to a landfill. The sludge, which contained ^{226}Ra and ^{228}Ra at about 0.9 kBq/kg (25 pCi/g) each, triggered portal radiation detectors at the landfill, and this initiated an investigation by the Pennsylvania Department of Environmental Protection (Rittiger and Yusko 1996).

As brine flows through pipes at the oil field, temperatures tend to drop and solutes tend to precipitate, forming a scale consisting of sulfates, carbonates, and silicates of calcium, strontium and barium along the interior walls. Radium tends to coprecipitate with these compounds, resulting in a radioactive scale. ^{226}Ra concentrations as high as 15,000 kBq/kg (400,000 pCi/g) have been reported, but typical concentrations are 4-400 kBq/kg (100-10,000 pCi/g). Exposure scenarios associated with these scales include gamma-ray exposure of workers at oil-production platforms and exposure to soil contamination at pipe-reaming facilities. Operations like the latter are conducted to maintain flow at the oil wells. Scale can be removed from pipes and other production equipment by mechanical methods, including cutting, shearing, and high-pressure blasting with water, sand and cryogenic carbon dioxide pellets (Lancée and others 1997). Chemical decontamination methods that use salts of amino carboxylic acids and proprietary reagents are available for the dissolution of scale and other surficial TENORM materials; radium can be precipitated from the spent solutions and the solid concentrate disposed of (Coll 1997; Lancée and others 1997).

Sludges are related deposits, typically found settled on the bottoms of equipment and storage tanks at various points in the oil-gas-water separation processing stream. The sludges are often oily, and disposal in burn pits used to be common. Large quantities of dewatered TENORM-contaminated scales and sludges have been stored in barrels at production facilities pending development of regulatory guidance. In 1992, an estimated 410,000 barrels of such TENORM waste was stored in Louisiana alone. In 1994, two commercial state-licensed TENORM-waste facilities opened in Louisiana and Texas; TENORM waste is diluted to reduce its specific activity to meet state criteria for reuse or disposal. Limited quantities of TENORM wastes from the oil and gas industry have been disposed of at low-level radioactive-waste sites licensed by the Nuclear Regulatory Commission (Minerals Management Service 1996a).

In addition to near-surface burial, TENORM waste can be disposed of by deep subsurface encapsulation in abandoned well bores or injection into permeable formations. Geologic and engineering criteria for such disposal on the Outer Continental Shelf of the Gulf of Mexico have recently been released by the Minerals Management Service (1996b). State regulations on down-hole encapsulation and injection of TENORM oil and gas wastes in onshore wells are in place or pending in Texas, Louisiana, and Mississippi (Minerals Management Service 1996a).

Equipment and piping that handle only the natural-gas fraction are not subject to scale and sludge deposits. However, radon-222 is carried from the reservoir with the gas, and its decay products tend to plate out on the interior surfaces of pipes, valves, and equipment in the gas plant (Gesell 1975). Short-lived, gamma-emitting radon decay products, such as bismuth-214, can pose an exposure hazard to plant workers, but the environmental fate of the longer-lived decay products ^{210}Pb and ^{210}Po is of concern after disposal of scrap metal from such operations, as is occupational exposure during maintenance and repair of disassembled equipment (Summerlin and Prichard 1985).

MUNICIPAL WATER TREATMENT

Conventional water-treatment processes designed to remove suspended solids and dissolved chemical contaminants from drinking-water supplies also remove radionuclides. During the period of atmospheric nuclear-weapons testing, the US Public Health Service and others did much work on removal of fission products, such as strontium-89 and strontium-90, from water supplies (Straub 1971). A variety of treatments, including lime-soda ash softening and phosphate coagulation, that were shown to be effective for radiostrontium removal can also remove substantial quantities of other alkaline earth metals, including radium (Menetrez and Watson 1983). Lime softening is effective in removing uranium from water. The radionuclide concentrations in the sludges generated by these treatments will be a function of the raw-water radionuclide concentrations and the radionuclide-removal efficiencies.

Regional geology is the key determinant of raw-water radionuclide concentrations. Water supplies with elevated concentrations of radium are found with the greatest frequency in the north central and Coastal Plain states. Water supplies with elevated levels of uranium are found most frequently in the western states (Horton 1985). Groundwater supplies with elevated concentrations of ^{210}Po have been reported in Florida (Harada and others 1989). Lime-softening sludges from water supplies in Illinois and Wisconsin that have raw-water ^{226}Ra concentrations of 0.04-0.2 Bq/L (1-5 pCi/L) have ^{226}Ra concentrations of 0.2-1.1 kBq/kg (6-30 pCi/g) (EPA 1993b). The sludges are most often disposed of in onsite lagoons or at municipal landfills with little regulatory control (for further discussion see chapter 7).

Treatments designed specifically to remove radium include coprecipitation with barium sulfate and selective sorbents. The latter include ion-exchange resins, barium sulfate-coated alumina, and manganese dioxide-coated polymers. Some of these can have ^{226}Ra concentrations as high as 3,700 kBq/kg (100,000 pCi/g) and might require disposal in low-level-waste burial grounds; likewise, brines from the regeneration of high-efficiency radium-removal resins might have high radium concentrations that present liquid-waste disposal problems. Indeed, at some municipal wastewater-treatment plants, elevated concentrations of radium in sewage sludge have been attributed to residual materials discharged to the sewer systems by drinking-water treatment plants (Nuclear Regulatory Commission 1997b). Activated-carbon filters, used for removal of the short-lived ^{222}Rn, can be handled with a delay-and-decay method before disposal (Lowry 1983).

METAL MINING AND PROCESSING

This category has by far the largest TENORM solid-waste volume—an estimated US inventory of about 50 billion tons—most of it with NORM concentrations less than 10 times background. On the basis of geologic reasoning, Bliss (1978) has outlined the types of metallic ores (other than uranium) whose mining and extraction might lead to TENORM problems. The list is broad and includes:

- Ores of rare-earth elements, molybdenum, gold, aluminum, lead-zinc, iron, tin, vanadium, copper, and other metals (commercial-scale byproduct recovery of uranium has occurred in connection with the extraction of copper and gold).

- Placer deposits of any metal (for thorium and its decay products).

- Ores that result from intense weathering, such as bauxite.

The remainder of this section focuses on metal resources, but selected non-metal resources might be associated with TENORM (Bliss 1978). These include organic deposits (such as black shales), fluorspar, granite, and clays.

Liquid and solid wastes from metal mining and processing include mine waters, overburden, mill tailings, pipe scales, smelter slags, and spent leachates. The presence of sulfide minerals in overburden and tailings is an important consideration. Oxidation of these materials on exposure to air generates sulfuric acid. As seen in chapter 2, one can expect migration of radium to differ from migration of uranium and thorium under such a weathering regime.

Although high sulfate concentrations in processing and disposal environments will limit the mobility of radium, the presence of other anions associated with metal-extraction processing can increase radium mobility. Chlorination is a process in which ores are treated with chlorine gas and then water to recover soluble metallic chloride salts; the process is used extensively with gold ores. At a plant in Oregon, chlorination of zircon-bearing sands was used to extract zirconium, niobium, tantalum, and hafnium. The process rendered radium, as well as these economic metals, water-soluble. The finely ground process tailings contained ^{226}Ra at about 20 kBq/kg (500 pCi/g), much of it occurring presumably as soluble $RaCl_2$. Seepage water at this tailings disposal site contained up to 1.7 kBq/L (45,000 pCi/L) (Boothe and others 1980; Bliss 1978).

GEOTHERMAL ENERGY PRODUCTION

TENORM wastes associated with geothermal-energy production are similar to those associated with oil and gas production: temperature changes lead to precipitation of solids from hot formation waters in piping, equipment, and retention ponds at the surface. ^{226}Ra and ^{228}Ra are the radionuclides of concern in the pipe scales and the solids dredged from holding ponds for spent geothermal fluids. The possibility of locally increased atmospheric ^{222}Rn concentrations near geothermal plants exists (Gesell and Adams 1975).

OTHER INDUSTRIES

Metal casting: Foundries use refractory sands to create molds for casting steel-alloy parts. The molds are eventually disposed of in landfills. The foundry sands—mined from deposits in Florida, South Africa, Australia, India, and Brazil—contain elevated concentrations of uranium and thorium that occur in heavy accessory minerals, such as zircon and monazite. State regulations restricting the disposal of radioactive material and the use of portal radiation monitors at landfills have in some cases made it difficult for the industry to dispose of discarded casting molds (Anonymous 1995).

The radon-emanation coefficients of these accessory minerals tend to be low, about 0.1-5% for zircon and monazite, compared with about 10-40% for soils and UMT (Landa 1987; Barretto and others 1975). Such differential environmental mobility factors should be considered for the atmospheric pathway in radiologic dose assessments of these wastes.

Pulp and paper: Radium-bearing barium sulfate scales have been found deposited at various points in paper mills (Coll 1997). Such scales probably were responsible for an incident reported by the Pennsylvania Department of Environmental Protection in which a paper-mill digester tank taken to a scrap-metal facility triggered radiation monitors (Yusko 1997).

Soils from former radium-processing or -manufacturing sites: Radium was extracted from uranium ores at a variety of sites in the United States during the early 20th century. The radium was used extensively for medical purposes and for the production of luminous paint until the 1950s. Residual radium contamination of soils at such sites has required cleanup under the Superfund or other programs (Neiheisel 1990; Simon 1990; Landa 1984).

TENORM in Selected Nonnuclear Industries in Other Nations: Most of the categories covered in the discussion above have been the subjects of multiple investigations of TENORM occurrences in the United States. Some additional categories of TENORM in nonnuclear industries have received more attention in other nations and are noted briefly in table 3.2.

Many industrial processes use feed materials with NORM or have TENORM as byproducts. In some cases, the existence of TENORM is ignored by industrial managers and workers, and TENORM wastes might yet be discovered. But some nonnuclear industries are aware of TENORM in their processes or wastes. Table 3.2 presents a selected list of nonnuclear industries in which TENORM play a role either in processes or as byproducts.

MINIMIZATION OF TENORM

Risks posed by TENORM can sometimes be reduced or redirected to other populations by the application of specific technologies. For example, in the case of radionuclide removal from municipal drinking-water supplies, worker exposure might increase because of handling of radionuclide-bearing treatment residues (such as sludges, spent ion-exchange resins, and spent granular activated carbon) or inhalation of emanated radon and its decay products; at the same time, exposure of the water-supply users will decrease. The use of scale inhibitors or in situ removal of radionuclides from oil-field production fluids by the introduction of sorbents downhole can limit the buildup of TENORM in piping and equipment (Lancée and others 1997). The removal of uranium as an economic product from phosphoric acid production circuits will decrease the exposure of people who obtain foodstuffs from fertilized soils.

Table 3.2. TENORM in Selected Nonnuclear Industries in Other Nations

Industry	Comment	Concentration	References
Fertilizer	Possibility of public exposure exceeding 1 mSv	Feed material with ^{238}U at 1.5 kBq/kg. Scales with ^{226}Ra and/or ^{228}Ra at 100 kBq/kg	(Paschoa 1997a; Janssens and Markkanen 1997)
Phosphoric acid	Gypsum is main byproduct	^{226}Ra at 1 kBq/kg in gypsum; ^{226}Ra at 100 kBq/kg in precipitate	(Janssens and Markkanen 1997)
Metal production and smelters	Tin, lead, bismuth, titanium (from ilmenite and rutile), aluminum (from bauxite), iron-niobium (from pyrochlore and columbite)	Wide range of ^{226}Ra and ^{228}Ra contamination	(Paschoa 1997a; Breas and van der Vaart 1997; Janssens and Markkanen 1997)
Rare earths, monazite sands	Bioavailability of radium after monazite sands are chemically processed should be considered	Monazite sands: ^{226}Ra at 0.06–6.7 kBq/kg; ^{228}Ra at 0.3–33 kBq/kg; ^{238}U at 3.6 kBq/kg; ^{232}Th at 0.52 kBq/kg Zircon sands: ^{238}U at 125 kBq/kg; ^{232}Th at 43 kBq/kg	(Paschoa 1997a, 1994; Carter and Coundouris 1993; Paschoa 1993; Spezzano 1993; Chhabra 1966)

Table 3.2. (continued)

Industry	Comment	Concentration	References
Coal-mine dewatering plants	Uranium and thorium concentrations can vary by factor of 10	Uranium and thorium at 0.2–10 kBq/kg	(Janssens and Markkanen 1997)
Pigments	Feed material from monazite separation	Feed material: ^{238}U and ^{232}Th at 1 kBq/kg each Wastes streams: ^{238}U and ^{232}Th at > 5 kBq/kg each	(Janssens and Markkanen 1997; Paschoa 1994, 1993)
Thoriated tungsten lamps, welding rods, gas mantles	High thorium in thoriated tungsten lamps, thoriated welding rods, ashes of gas mantles	Thoriated lamps and rods: ^{232}Th at 500 kBq/kg. Ashes of gas mantles: ^{232}Th at 150 kBq/kg	(Janssens and Markkanen 1997; Cullen and Paschoa 1978)
Natural stones and shales	Granite and shales are found in several parts of world	Thorium, uranium, ^{40}K typically at 1–5 kBq/kg	(Janssens and Markkanen 1997)
Petroleum	Wide range of radium concentrations, but higher ones are in scales	^{226}Ra typically at 1–100 kBq/kg; Highest reported: 4×10^3 kBq ^{226}Ra/kg	(Paschoa 1997b, 1997a; Janssens and Markkanen 1997; Heaton and Lambley 1995; Kolb and Wojik 1985)

NONRADIOLOGIC IMPACTS

This report focuses on the radionuclides associated with TENORM. But nonradioactive inorganic and organic contaminants are also associated with these materials. For example, metal-mining wastes can contain a wide variety of inorganic contaminants associated with the ores and their processing; sludges with elevated radium concentrations at oil-water separators can also contain appreciable concentrations of oil. In establishing groundwater standards for remedial actions at inactive uranium-processing sites (40 CFR Part 192), EPA provided specific concentration limits for nitrate and molybdenum, as well as for uranium and radium, because these constituents had been found in high concentrations at many UMT sites (EPA 1995). Risk assessments for TENORM should consider such exposures.

CONCLUSIONS

TENORM present unique problems because of their large volumes and widespread occurrence in industrial products, byproducts, and wastes. The physical, chemical, and radiologic properties of TENORM vary widely. ^{226}Ra and its decay products are the radionuclides of primary concern, but other uranium- and thorium-series nuclides should also be considered. As discussed further in chapter 4, the leachability, sorption, and biologic availability of these radionuclides can be expected to vary with the processing history and siting environment of the TENORM. We might not know all sources of TENORM.

4
Role of Exposure and Dose or Risk Assessments in Developing Radiation Standards

This chapter discusses the use of exposure and dose or risk assessments in providing a technical basis for standards for radionuclides in the environment. It does not discuss in detail the current approaches to exposure and dose or risk assessments of radionuclides, such as their use in demonstrating compliance with standards.

The standards for radionuclides in the environment discussed in this report are directed only at protection of humans. According to available data and modeling, radiation standards that would provide acceptable protection of humans would ensure that other species are not put at unreasonable risk, although individual members of a species occasionally might be harmed (IAEA 1992; ICRP 1991). Risks to biota other than humans and the issue of whether separate standards are needed for their protection are not considered in this study.

The chapter begins with a general discussion of the risk-assessment process for carcinogens and the application of the process to radionuclides. That is followed by a discussion of the calculational elements of dose or risk assessments for radionuclides in the environment and the use of such assessments in developing standards. The chapter ends with the committee's views on suitable approaches to exposure and dose or risk assessments for purposes of developing standards for radionuclides in the environment.

RISK ASSESSMENT OF CARCINOGENS

As described, for example, by the National Research Council (1983) and the Office of Science and Technology Policy (OSTP 1985) the process of risk assessment of carcinogens, including radionuclides, generally consists of the following four components:

- Hazard identification.
- Estimation of dose-response relationship.
- Exposure assessment.
- Risk characterization.

Brief descriptions of these components and the usual approaches to addressing them for radionuclides are presented in the following sections.

Hazard Identification

Hazard identification for carcinogens entails a qualitative evaluation of the data bearing on an agent's ability to produce carcinogenic effects and the relevance of this information for humans. This component of the risk-assessment process is particularly important for chemical carcinogens because data on possible carcinogenic effects on humans often are lacking. However, hazard identification is not an important concern for radionuclides, including naturally occurring radionuclides found in TENORM, because widespread epidemiologic data have shown conclusively that ionizing radiation can cause cancer in humans (for example, National Research Council 1990; 1988).

Noncarcinogenic health effects also are potentially important for some radionuclides that can be found in TENORM. Particularly for uranium, chemical toxicity in the kidney after exposures above some threshold is clearly demonstrated by a large amount of animal and human data (for example, Leggett 1989). For uranium, an important issue is whether current radiation standards for the public would prevent chemical toxicity in the kidney. Resolution of that issue depends on a number of factors including: the assumed threshold for chemical toxicity in the kidney; selection of an appropriate safety factor below the assumed threshold for protection of public health, including protection of unusually sensitive populations (such as infants and children); and the relationship between kidney burden and radiation dose from uranium. On the basis of consideration of those factors and current data and models, Kocher (1989) concluded that chemical toxicity generally should be considered in developing health-protection standards for the public with respect to ingestion and inhalation of soluble or insoluble uranium. For example, the Environmental Protection Agency has based a proposed standard for uranium in drinking water on prevention of chemical toxicity in the kidney, as well as limitation of radiation dose (see chapter 7). However, the chemical toxicity of natural uranium appears to be potentially important only if the limit on radiation dose from exposure to uranium is greater than about 0.25 mSv (25 mrem) per year and if the dose from uranium results primarily from ingestion and inhalation rather than external exposure (Kocher 1989).

The chemical toxicity of naturally occurring radionuclides found in TENORM and its effect on setting radiation standards is not considered further in this report.

Estimation of Dose-Response Relationship

Estimation of the dose-response relationship for carcinogens is the process of estimating the magnitude, or an upper bound on the magnitude, of the carcinogenic effect of any given dose. For radiation exposure, best estimates of the carcinogenic effect of a given dose normally are emphasized.

Estimation of the carcinogenic response to exposure to particular radionuclides is greatly facilitated by the generally held view (which is based on observation and modeling) that the absorbed dose in tissue is the fundamental physical quantity that determines the response to any exposure. The absorbed energy in tissue depends only on the radiation type and its energy (NCRP 1993a; ICRP 1991), not on the source of the radiation. Therefore, in contrast with the current situation for chemical carcinogens, animal or human studies to estimate the dose-response relationship for exposure to every radionuclide of potential concern are not required. Rather, the dose-response relationships developed for different types of radiation (such as photons, electrons, and alpha particles) can be applied to exposures to any radionuclide once the energies and intensities of all the kinds of radiation emitted in the decay of the radionuclide are known.

Current estimates of the dose-response relationships for radiation exposure obtained, for example, by the National Research Council's Committee on the Biological Effects of Ionizing Radiations (National Research Council 1990; 1988) are based on studies of human populations that received radiation doses considerably above environmental levels. The most important study groups include the Japanese atomic-bomb survivors, who received high doses from photon exposures of all organs and tissues of the body, and various groups of miners, who received high doses to the lung from exposure to alpha particles emitted by the short-lived decay products of radon.

Applying the dose-response relationships for radiation exposure estimated by such groups as the BEIR committee to the considerably lower doses of concern in controlling routine exposures of the public requires assumptions about extrapolation to doses and dose rates beyond the range of direct observation (ICRP 1991). For purposes of risk management, the carcinogenic response to environmental radiation generally is assumed to be a linear function of dose without a threshold. Although the assumption has often been challenged, the consensus among regulatory authorities and expert organizations is that there is insufficient evidence to support a departure from the linear, no-threshold dose-response relationship for regulatory purposes. This issue has not been investigated in the present study of standards for TENORM.

Exposure Assessment

Exposure assessment is the process of identifying a group or groups of individuals who might be at risk and describing for them—on the basis of assumptions, observations, or modeling—the various routes (pathways) of exposure and their magnitude, frequency, and duration. Exposure assessment is essentially the same for radionuclides and hazardous chemicals except that external (direct) exposure to penetrating radiation (such as photons) is an important pathway for radionuclides but not for hazardous chemicals.

The process of exposure assessment for radionuclides in the environment is considered in more detail later in this chapter.

Risk Characterization

Risk characterization is the process of combining the information obtained from hazard identification, estimation of the dose-response relationship, and exposure assessment to describe the carcinogenic risk associated with expected or assumed human exposures to the agent of interest.

Risk characterization for radiation exposure is relatively straightforward and is almost always expressed in terms of the numerical probability of cancer induction in an individual or the number of cancers in a population over a defined time period. Some approaches to risk characterization for radiation exposure focus on fatal cancer as the end point of concern, whereas others focus on cancer incidence; the difference between the two is not important for most organs and tissues at risk (ICRP 1991). Other issues in risk characterization include the risks that might be experienced by particularly susceptible populations and alternative approaches to expressing risk. Those issues have not been considered in this study.

ELEMENTS OF RADIATION RISK ASSESSMENT

This section discusses the calculational elements that normally make up an assessment (estimation) of risk posed by exposure to radionuclides in the environment. It does not apply to risk assessments for exposure to radon and its short-lived decay products in air. As noted in chapter 1, risk assessments of radon usually are based directly on epidemiologic data without the need to estimate dose from a given exposure and risk per unit dose.

There is some overlap between this discussion and the general discussion of risk assessment for carcinogens above. However, we are discussing radiation risk assessment separately mainly because the approaches to estimating radiation risks were developed long before the general risk-assessment process for carcinogens was formally laid out. Radiation risk assessment usually includes estimates of dose from a given exposure, which are generally not part of risk assessments of chemical carcinogens.

An exposure and dose or risk assessment of radionuclides in the environment addresses the relationship between the concentrations of radionuclides in various media (air, water, soil, or other materials) at particular locations and the resulting radiation doses or risks to exposed individuals or populations. Knowledge of this relationship can be used either to derive estimates of dose or risk corresponding to given concentrations of radionuclides in the environment or to derive estimates of concentrations of radionuclides in the environment corresponding to a given dose or risk.

As shown in figure 4.1, an assessment of risk corresponding to given concentrations of radionuclides in the environment, or vice versa, consists of three elements:

- An assessment of internal and external exposures, which provides estimates of intakes of radionuclides by ingestion and inhalation, and estimates of exposures to external radiation per unit concentration of particular radionuclides in environmental media.

- An estimate of radiation dose, given the estimates of radionuclide intakes and external exposures.

- An estimate of risk, given the estimate of dose from internal and external exposures.

As noted previously, the measure of risk posed by radiation exposure normally is assumed to be cancer mortality, although cancer incidence (morbidity) is calculated in some cases. In many environmental-radiation assessments, the desired end point for the calculations is dose, rather than cancer risk, especially when an assessment is used to develop a radiation standard expressed in terms of dose or to demonstrate compliance with a dose standard. When dose is the desired end point, only the first two elements listed above are included in the assessment. Furthermore, as noted previously, dose often is calculated as an intermediate step even when risk is the desired end point for the assessment.

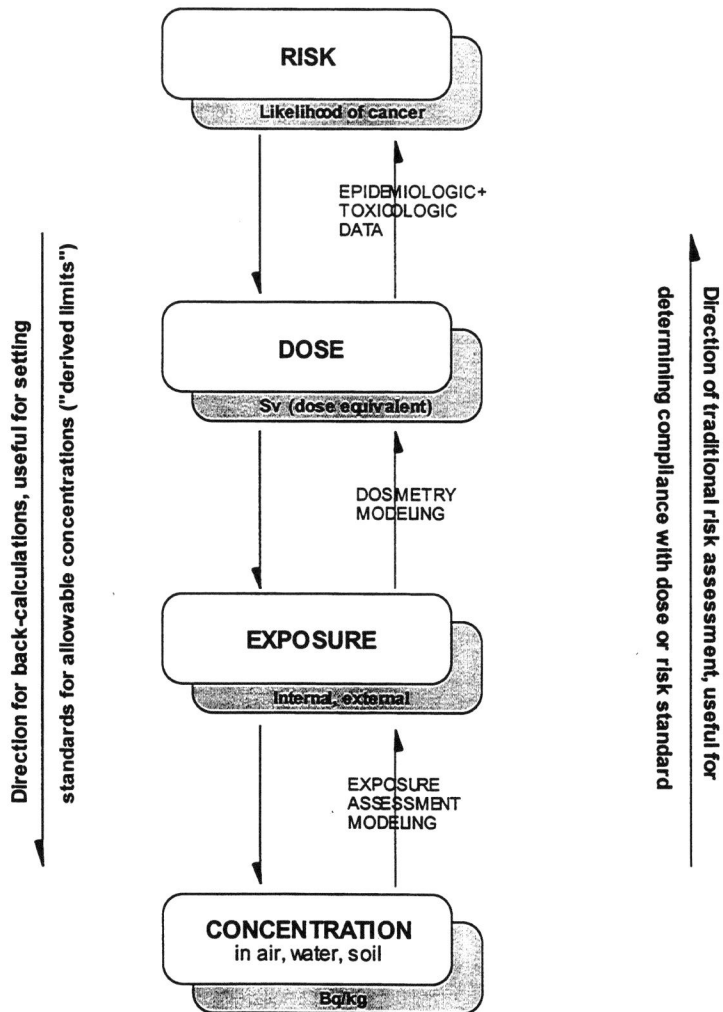

Figure 4.1. Relationships among risk, dose, exposure, and concentration of radionuclides in environmental media. Typical example units are shown.

The approach to estimating dose and risk for given internal and external exposures (the second and third elements of a risk assessment listed above) often is straightforward. Estimates of committed doses from internal exposure per unit activity intakes of radionuclides by ingestion and inhalation can be obtained, for example, from federal guidance (Eckerman and others 1988). Similarly, for radionuclides dispersed in air, water, or soil, estimates of external dose rates per unit activity concentration are given in federal guidance (Eckerman and Ryman 1993). Estimates of dose from internal and external exposure then can be converted to estimates of risk by using the nominal fatal-cancer risk per unit dose discussed in chapter 7. These simplified approaches to estimating dose and risk generally are adequate for purposes of controlling radiation exposures (that is, in developing standards and in demonstrating compliance with standards), but they might not be adequate for estimating doses and risks for real exposure situations (see chapter 11).

The approach to estimating internal and external exposures per unit concentration of radionuclides in various environmental media (the first element of a risk assessment) generally is more complicated than the approach to dose and risk estimation described above, especially for internal exposure. Estimates of ingestion intakes of radionuclides can require estimates of transfers of radionuclides through various terrestrial or aquatic food-chain pathways to humans, including transfers into food crops, meat, milk, and fish; and estimates of inhalation intakes can require estimates of transfers of radionuclides from various environmental media, such as soil, into the air. Exposure assessment also requires estimates of so-called usage factors, including intakes of contaminated water, air, soil, foodstuffs, or other materials by humans or livestock and residence times and shielding factors for external exposure.

The general approach to exposure assessment for radionuclides in the environment is depicted in figure 4.2. The source compartments can include water, air, soil, or other materials (such as contaminated buildings or equipment), and the exposure compartments can include the source compartments plus various foodstuffs or other materials into which radioactivity is transferred. The particular exposure pathways that should be considered in any assessment depend on such factors as the characteristics of the site, the source compartments of concern, the physical and chemical forms of radionuclides, and the assumptions about living habits of exposed individuals or populations. Examples of important exposure pathways for different source compartments are as follows.

For contaminated soil, important exposure pathways generally include:

- Consumption of vegetables, fruits, and grains grown in the contaminated soil.

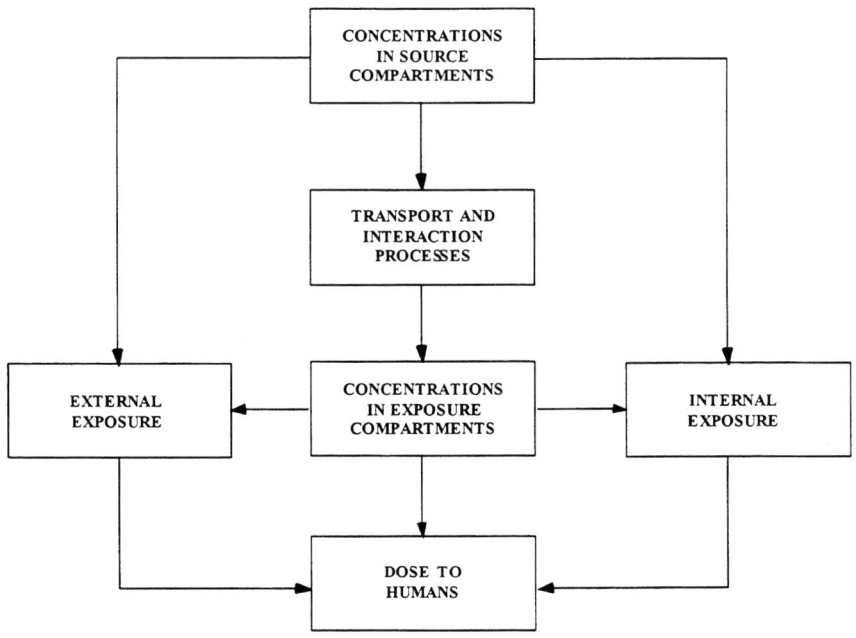

Figure 4.2. Relationships between concentrations of radionuclides in source compartment and resulting doses to humans from internal and external exposure.

- Consumption of milk and meat obtained from livestock that eat pasture grass grown in the contaminated soil.
- Direct consumption of the contaminated soil.
- Inhalation of radionuclides from contaminated surface soil suspended in the air.
- External exposure to the contaminated soil.

For contaminated groundwater, additional important pathways of internal exposure generally include:

- Direct ingestion of water from the contaminated source.
- Use of the contaminated source as a water supply for agricultural purposes, such as watering of livestock; irrigation of vegetables, fruits, and grains; and irrigation of pasture grass consumed by livestock.

For contaminated surface water, additional important exposure pathways generally include:

- The ingestion pathways listed above for contaminated groundwater.
- Consumption of aquatic foodstuffs (such as fish) obtained from the contaminated source.
- External exposure to the contaminated source during such activities as swimming, boating, and residence along contaminated shorelines.

For contaminated air, important exposure pathways generally include:

- Inhalation and external exposure to airborne radionuclides in the source compartment.
- Exposure to radionuclides deposited on vegetation and the ground surface, including external exposure and ingestion of radionuclides incorporated into foodstuffs (vegetables, fruits, grains, meat, and milk).
- Inhalation of deposited radionuclides that are resuspended in the air.

Finally, if the source compartment consists of contaminated structures or equipment, credible scenarios for ingestion, inhalation, and external exposure would need to be developed and evaluated on the basis of the characteristics of the particular exposure situation of concern.

Most models for terrestrial and aquatic food-chain pathways assume that the concentration of a radionuclide in the exposure compartment of interest (such as milk) is a constant multiple of the concentration in the source compartment (such as soil). Thus, food-chain pathway models generally use radionuclide-specific transfer factors (also called concentration factors or concentration ratios in some cases). For pathways involving exposure to contaminated soil, the long-term retention of radionuclides in surface soil, which depends on the solid-solution distribution coefficient (K_d) for the radionuclides, also is important. The models and parameter values for the various internal and external exposure pathways of concern in exposure assessments for radionuclides in the environment are discussed in a number of references (for example, IAEA 1996b, 1994; NCRP 1984d; Till and Meyer 1983).

Additional issues are particularly important with respect to TENORM. Soluble radionuclides are more available for biologic uptake than those sorbed on soils or sediments, so the partitioning of radionuclides among these phases—between groundwater and aquifer materials; between river water and suspended or bottom sediments—is important in pathway modeling. A laboratory-derived distribution coefficient, K_d, is typically used, where

$$K_d = \frac{\text{concentration of radionuclide sorbed on sediment}}{\text{concentration of radionuclide remaining in solution}}.$$

For a given element, the coefficient can be expected to vary with the chemical speciation of the element, the solute chemistry of the water, the mineralogy and surface area of the solids, redox conditions, and pH. For example, the presence of competing ions in solution can decrease the sorption of a radionuclide. That effect was seen in the sorption of ^{226}Ra from a sodium chloride oil-production brine by soils and marsh sediments where the percentage of radium sorbed increased with brine dilution (Landa and Reid 1983). In groundwater, uranium partitioning to the solid phase can be expected to increase with more-reducing conditions along a flowpath. Laboratory-derived K_d values should attempt to simulate field conditions at the TENORM site, and generic, literature-derived K_d values should be used with caution.

Much of the nutrients locked up in unweathered rock fragments might not be available for plant uptake; the chemical forms of radionuclides in TENORM might greatly influence their environmental mobilities and biologic availabilities. Total radionuclide concentrations in TENORM might not be reflective of the biologically-labile pool, and some subsegment of the total could be of greater value in assessing biologic uptake and resulting dose. For example, in assessing the hazard due to inhalation of radon decay products, radon that does not escape the radium-bearing mineral matrix is not a concern. The radon emanation coefficient is the fraction of the radon formed in a radium bearing solid that escapes to the atmosphere. It can vary widely in natural and industrial materials; for example, although ^{226}Ra coprecipitated with barium sulfate generally emanates less than 1% of the ^{222}Rn generated, that coprecipitated with or sorbed on iron hydrous oxide can emanate nearly 100% (Hahn 1936). For TENORM with high emanation coefficients, radon release and radon decay-product inhalation might be the pathway of concern. For TENORM with low emanation coefficients, whole-body gamma exposure might yield the limiting dose.

For the inhalation or ingestion pathway, the solubility of the TENORM particles in bodily fluids (such as lung serum for inhaled particles, or stomach acid) and in the soil and aquatic environments where TENORM resides can exert a major influence on the direct and indirect uptake of radionuclides by humans. In vivo solubility can be affected by several factors such as particle size, position of nuclides in less-accessible particle interiors versus those on particle surfaces (for example, heat-volatilized ^{210}Pb/^{210}Po condensed onto cooling particle surfaces in flue dust of thermal-process phosphate plants and in coal-combustion fly ash), and kinetics. In soil and aquatic environments, radionuclide solubility can vary with pH, presence of complexing or precipitating ions in contact solutions, and redox conditions. Plant uptake of radionuclides differs between different members of a decay series and between species for a given nuclide; the biologic uptake of radium by terrestrial plants has been reviewed by Simon and Ibrahim (1990). If biologic-uptake data are not available, leaching tests with dilute acids to assess readily dissolvable fractions, or salt solutions to assess ion-exchangeable fractions can be used to provide an index of biologic availability.

USE OF DOSE OR RISK ASSESSMENT IN DEVELOPING RADIATION STANDARDS

In principle, radiation standards expressed in terms of dose or risk can be developed without the need to consider exposure-pathway and dose or risk assessments for the exposure situations of concern. If a standard is expressed directly in terms of risk, all that is required in developing the standard is a judgment about a limit on acceptable risk. Similarly, if a standard is based only on a judgment about acceptable risk but is expressed in terms of dose as a surrogate for risk, the only additional requirement is an assumption about the risk per unit dose. For purposes of developing standards expressed in terms of dose, the risk per unit dose normally is assumed to be independent of the particular radionuclides and exposure pathways of concern.

In practice, however, some type of exposure-pathway and dose or risk assessment normally is used in developing radiation standards, even when the standards are expressed only in terms of dose or risk. For example, such assessments are required by the National Environmental Policy Act whenever imposition of a standard would have substantial economic or environmental impacts. More generally, dose or risk assessments are important in developing radiation standards because risks posed by environmental exposures are not directly observable and dose often cannot be measured, especially the dose from internal exposure. Dose or risk assessments are important in developing radiation standards in the following ways.

First, if a standard is expressed in terms of dose or risk and is based on a judgment about acceptable risk, as described above and discussed further in chapter 5, a dose or risk assessment for the exposure situations of concern can be used to demonstrate whether compliance with the standard is reasonably achievable. Such demonstrations help in gaining acceptance of the standard by the public and affected parties.

Second, if a standard is expressed in terms of dose or risk but is based on a judgment about the achievability of doses or risks for the exposure situations of concern, as described above and discussed further in chapter 5, a dose or risk assessment clearly is required to support the conclusion that the standard is reasonably achievable. In making such a judgment, some type of cost-benefit analysis often is performed in which the costs of achieving various doses or risks are compared with the corresponding benefits in health risks averted in exposed populations (for example, Wolbarst and others 1996; ICRP 1989b).

Third, radiation standards might be expressed in terms of directly measurable quantities, such as concentrations of radionuclides in environmental media or external exposure rates, rather than in terms of dose or risk which cannot be measured. Standards expressed in terms of measurable quantities often are developed to facilitate demonstrations of compliance, especially in the case of standards for naturally occurring radionuclides (see chapter 7). If such a standard is based on an assumed limit on dose or risk, exposure and dose or risk assessments clearly are required in deriving the limits on measurable quantities.

However, an exposure and dose or risk assessment is not necessarily required even in developing environmental standards expressed in terms of directly measurable quantities. Consider, for example, current standards for alpha-emitting radionuclides in drinking water, radioactivity in liquid discharges from particular mines or mills, control and cleanup of residual radioactive materials at uranium and thorium mill-tailings sites, and indoor radon, all of which are concerned with exposures to naturally occurring radionuclides (see chapter 7). In each standard, the quantitative criteria are expressed in terms of measurable quantities, and the criteria were developed on the basis of considerations of levels of radioactivity that are reasonably achievable for the exposure situations of concern, given existing background levels and the ability of current technologies to reduce them. However, the cost-benefit analysis used in developing the criteria in each case was performed with respect to levels of radioactivity, rather than with respect to the corresponding doses or risks to individuals or populations. Thus, the values of measurable quantities that were judged reasonably achievable were derived essentially without concern for the resulting doses or risks. Those examples illustrate that an exposure-pathway and dose or risk assessment is required in developing standards expressed in terms of measurable quantities only when the standards are based on a judgment about acceptable dose or risk.

Additional discussions on the various ways that standards for TENORM might be expressed and the implications of the different forms of standards for exposure-pathway and dose or risk assessment are presented in chapter 11.

SUITABLE APPROACHES TO RISK ASSESSMENT IN DEVELOPING STANDARDS

As indicated earlier, it is not the purpose of this chapter to discuss in detail the kinds of exposure and dose or risk assessments that might be used in developing a technical basis for radiation standards, especially standards for TENORM. However, the committee offers the following general observations on suitable approaches to risk assessment in developing radiation standards.

First, in developing radiation standards, it is appropriate to use stylized methods of exposure and dose or risk assessment for assumed reference conditions, provided that the assumed conditions are reasonably representative of the exposure situations of concern and that the regulations permit the use of alternative and more realistic approaches for specific exposure situations, especially in demonstrating compliance with the standards. The use of stylized methods of exposure, dose, and risk assessment in developing environmental standards for radionuclides is consistent with the approach used in developing secondary limits (limits on intakes of radionuclides or concentrations of radionuclides in air) for control of radiation exposures in the workplace (for example, Eckerman and others 1988).

Second, exposure and dose or risk assessments used in developing standards should be reasonably realistic, particularly the assumptions about pathways for internal and external exposure, including transfers of radionuclides among various environmental compartments, transfers through terrestrial and aquatic food chains, and the various usage factors for internal and external exposure pathways. That is, the assumptions should not be intended to greatly overestimate or underestimate actual outcomes for the exposure situations of concern. The need for reasonably realistic assessments is particularly important if standards are based on cost-benefit analyses with respect to dose or risk. A dose or risk assessment that is unreasonably conservative or nonconservative could lead to standards that either are not reasonably achievable or are not adequately protective of public health.

5
Basic Approaches to Regulating Radiation Exposures of the Public

This chapter presents a general discussion of basic approaches to regulating exposures of the public to radionuclides in the environment. These approaches are applied without regard for whether the radionuclides of concern are naturally occurring or human-made. The primary purpose of this discussion is to provide information that would be useful in understanding the many guidances and regulations regarding radionuclides in the environment discussed in chapters 7-10. The guidances and regulations cited here as examples are discussed in more detail in chapter 7.

The fundamental purpose of any standard for radionuclides in the environment is to limit health risks to exposed individuals and populations. Many standards for controlling radiation exposures of the public have been developed by the Environmental Protection Agency (EPA) and other federal agencies with responsibilities in radiation protection of the public, including the Nuclear Regulatory Commission and the Department of Energy. Furthermore, the EPA's guidances and regulations have been developed under the authority of various environmental laws that mandate different approaches to health protection of the public (Overy and Richardson 1995). Nonetheless, any standard is based on considerations of one of the following factors for the particular exposure situations of concern:

- Judgments about whether particular magnitudes of health risk to the public are *acceptable*.

- Judgments about whether particular magnitudes of health risk to the public are *achievable*.

In judging the acceptability of risks, radiation dose often is used as a surrogate for risk. Similarly, judgments about the achievability of risks might be represented by judgments about the achievability of releases of radionuclides to the environment, amounts of radioactivity in the environment, exposures, or doses.

JUDGMENTS ABOUT ACCEPTABILITY OF HEALTH RISKS

An important example of the use of judgments about acceptable health risks to the public in developing standards for control of radiation exposures is found in EPA's proposed federal guidance on radiation protection of the public (EPA 1994d) discussed in chapter 7. This guidance would apply to all controlled sources of exposure, including sources not associated with operations of the nuclear fuel cycle, but excepting exposures to indoor radon and beneficial medical exposures. Because the guidance would apply to nearly all sources of exposure other than natural background, it embodies important statements of principle about acceptable health risks to the public posed by radiation exposure.

EPA's proposed federal guidance on radiation protection of the public is based in large part on the traditional approach to radiation protection embodied in recommendations of the International Commission on Radiological Protection (ICRP 1991; 1977) and the National Council on Radiation Protection and Measurements (NCRP 1993a; 1987c). An important element of these recommendations, which is incorporated in EPA's proposed federal guidance, is the specification of a maximum allowable individual dose from exposure to all controlled sources combined. The specified primary dose limit for chronic exposure of individual members of the public is 1 mSv (100 mrem) per year.

The primary dose limit of 1 mSv (100 mrem) per year for all controlled sources combined was originally developed on the basis of considerations of acceptable risk. Specifically, on the basis of a review of other involuntary risks that the public has accepted in everyday life, ICRP (ICRP 1977) judged that an increase in lifetime risk of about 10^{-3} was an upper bound on acceptable risk posed by radiation exposure. The 1-mSv (100-mrem) limit on annual dose then was derived from the upper bound on acceptable risk by assuming a risk of fatal cancers of about 10^{-5} per millisievert (10^{-7} per millirem)[3] and continuous

---

[3] The current estimate of the risk per unit dose and its effect on the primary dose limit for individual members of the public is discussed in chapter 7.

exposure over a 70-y lifetime (ICRP 1977). Consideration of whether the primary dose limit was achievable was not an important factor in establishing its value. However, on the basis of experiences with the nuclear industry, ICRP expected that individual doses substantially below the limit were easily achievable for nearly all controlled sources.

A second example of the use of judgments about an acceptable risk in developing standards for controlling radiation exposures of the public is found in EPA's current regulations for airborne emissions of radionuclides developed under authority of the Clean Air Act and discussed in chapter 7. In contrast with the proposed federal guidance on radiation protection of the public discussed above, the regulations for airborne emissions apply only to particular sources (that is, not all airborne releases of radionuclides are subject to the regulations) and only to a single pathway of release to the environment.

In response to a lawsuit over standards for airborne emissions of vinyl chloride, the court of appeals mandated that standards for airborne emissions of hazardous air pollutants developed under the Clean Air Act must be based on a determination by EPA of a safe or acceptable level of risk to individuals or populations and an ample margin of safety below that level for protection of public health, but that the standards could not be based on considerations of technical feasibility or cost (for example, EPA 1989d). In response to the court's mandate, EPA established standards for airborne emissions of radionuclides (and other carcinogens) that would ensure that the lifetime cancer risk would not exceed about 10^{-4} for maximally exposed individuals and about 10^{-6} for average exposures in the population.

A third example of the use of judgments about an acceptable risk in developing standards for controlling exposures of the public is found in EPA's current regulations for remediation of contaminated sites; the regulations were developed under authority of the Comprehensive Environmental Response, Compensation, and Liability Act (CERCLA) and are discussed in chapter 7. The regulations, which apply to radionuclides and hazardous chemicals, specify that a lifetime cancer risk of about 10^{-4} is one of the criteria to be used in defining acceptable risk when considering the need for site remediation. However, the individual risk criterion of 10^{-4} in standards for remediation of contaminated sites under CERCLA is fundamentally different in concept from the upper bound on acceptable risk corresponding to the primary dose limit of 1 mSv (100 mrem) per year in the proposed federal guidance on radiation protection of the public and the limit on individual risk of 10^{-4} embodied in regulations for airborne emissions of radionuclides. The primary dose limit and the regulations for airborne emissions define limits on acceptable risk for the exposure situations to which they apply, but the risk criterion for remediation of contaminated sites defines a goal for acceptable cleanups that can be relaxed on the basis of many considerations, including a determination that achieving the goal is not cost-effective. In essence, the risk goal of 10^{-4} for remediation of contaminated sites defines an upper bound on negligible (trivial) risk for a particular exposure situation, but it does not define a limit on risk that must be met regardless of other circumstances.

Thus, several EPA guidances and regulations for radionuclides in the environment were based primarily on a priori judgments about acceptable health risks to the public for the exposure situations of concern. However, it would be misleading to assume that judgments about acceptable risk were the only basis for those standards. The proposed federal guidance on radiation protection of the public includes additional provisions, besides the primary dose limit of 1 mSv (100 mrem) per year for all controlled sources combined, that are based on considerations other than a limit on acceptable risk. The standards for airborne emissions of radionuclides also were cognizant of doses and releases that were reasonably achievable, and the cancer-risk criterion for cleanup of contaminated sites under CERCLA is only a goal that can be exceeded in acceptable cleanups under many circumstances (see chapter 7).

JUDGMENTS ABOUT ACHIEVABILITY OF HEALTH RISKS

Many EPA regulations that apply to specific sources or practices were based primarily on judgments about the ability of available technologies to control or reduce releases of radionuclides to the environment, levels of radioactivity in the environment, exposures, or doses, rather than a priori judgments about acceptable risks for the exposure situations of concern. Regulations based primarily on judgments about the ability of technologies to control or reduce radiation exposures of the public include standards for operations of uranium fuel-cycle facilities, radioactivity in community drinking-water systems, radioactivity in liquid discharges from particular mines or mills, uranium and thorium mill tailings, and management and disposal of spent fuel, high-level waste, and transuranic waste (see chapter 7). The federal guidance on indoor radon (see chapters 7 and 8) also was based in part on considerations of the cost effectiveness of available technologies for reducing existing concentrations of radon in homes. Finally, most cleanup decisions regarding contaminated sites subject to remediation under CERCLA have been based primarily on considerations of cost and feasibility, rather than compliance with the cancer-risk goal described in the previous section (see chapter 7). Thus, in controlling radiation exposures of the public, the standards for airborne emissions of radionuclides developed under the Clean Air Act are the only EPA regulations for specific sources or practices that were based primarily on judgments about acceptable risks.

A second important element of ICRP (1991) and NCRP (1993a) recommendations on radiation protection, which also is incorporated in the proposed federal guidance on radiation protection of the public (EPA 1994d), is the principle that exposures of individuals and populations should be as low as reasonably achievable (ALARA). Thus, compliance with the primary dose limit of 1 mSv (100 mrem) per year for all controlled sources combined does not, by itself, provide acceptable radiation protection of the public, because doses also should be reduced as far below the primary dose limit as practicable. Indeed, as discussed in chapter 7, application of the ALARA objective often is the most important consideration in radiation protection of the public (for example, NCRP 1993a).

In essence, all EPA regulations or guidances listed above that were based primarily on judgments about the achievability of releases of radionuclides to the environment, levels of radioactivity in the environment, exposures, or doses (and, therefore, risks) constitute an application of the ALARA objective to standard-setting itself. The standards for particular sources or practices represent risks that EPA judged reasonably achievable at any site, that is, the standards specify minimally acceptable performance for the particular sources or practices based on ALARA considerations. In each case, however, the judgments necessarily were somewhat subjective; there are no purely objective criteria for judging what is reasonably achievable, and judgments about the achievability of particular doses or risks for a particular exposure situation can be influenced by their magnitude in relation to doses or risks that have been judged acceptable for other exposure situations.

OTHER CONSIDERATIONS IN DEVELOPING STANDARDS

As described in the previous two sections, judgments about the acceptability of risks or the achievability of risks are fundamental to the development of all standards for controlling radiation exposures of the public. However, other considerations also can be important in developing radiation standards for the public, including the justification of practices, the ability to measure radioactivity in the environment at the levels of concern, and levels of natural background radiation.

Justification of Practices

A third important element of ICRP (1991) and NCRP (1993a) recommendations on radiation protection, which also is incorporated in the proposed federal guidance on radiation protection of the public (EPA 1994d), is the principle that all exposures should be justified, that is, that the benefits to society from any practice that increases radiation exposure should outweigh the overall societal costs. Justification of exposures is not an important concern for any of the guidances or regulations discussed in chapter 7, either because exposures cannot be avoided, as is the case with some exposures to naturally occurring radionuclides, or because a societal decision was made that operations of the nuclear fuel cycle for peaceful and defense purposes were, on the whole, beneficial.

In some cases, however, the principle of justification has been important in radiation protection of the public. For example, frivolous uses of radioactive materials (such as incorporation into children's toys or costume jewelry) have been banned by the Nuclear Regulatory Commission without regard for the magnitude of the resulting doses and risks because there is no overriding benefit.

Measurability of Radioactivity in the Environment

A standard for controlling radiation exposures of the public is useful in practice only to the extent that compliance with the standard can be verified by environmental measurements, including measurements of external radiation or quantities of radionuclides in air, water, soil, foodstuffs, or other materials. Therefore, the ability to measure radioactivity in the environment at particular levels can be an important consideration in developing standards for specific sources or practices. For example, current regulations for operations of uranium fuel-cycle facilities, radium and gross alpha-particle activity in community drinking-water supplies, radium in soil at uranium and thorium mill-tailings sites, and management and storage of spent fuel, high-level waste, and transuranic waste all took into account, to some extent, the ability to measure radioactivity in the environment at levels corresponding to the specified standards.

Although the ability to measure radioactivity in the environment is important, the measurability of various environmental surrogates for risk generally cannot be the primary basis for the standards. The difficulty with basing standards only on considerations of measurability is that the levels at which particular radionuclides can be measured in the environment do not correlate well with doses and risks to the public; furthermore, radionuclides posing the greatest risks might be the most difficult to measure at environmental levels of concern. For example, radionuclides that are high-energy photon emitters are the easiest to measure at low levels in the environment but often are relatively short-lived and so might result in relatively low doses and risks, whereas long-lived beta- or alpha-emitting radionuclides might be difficult to measure at low levels but could result in relatively high doses and risks. An additional consideration here is that the environmental levels of concern should be readily measurable with conventional and cost-effective techniques.

Natural Background Radiation

Radiation standards for the public generally have been developed in full recognition of the magnitude and variability of natural background radiation. The ubiquitous background of natural radiation, with its attendant and largely unavoidable doses and risks, has influenced the development of radiation standards in two ways.

First, geographic variations in natural background, excluding the large variations in levels of radon in homes, provide a perspective on whether the primary dose limit for the public of 1 mSv (100 mrem) per year for all controlled sources combined discussed earlier is reasonable (NCRP 1993a; ICRP 1991). In particular, it is regarded as unreasonable to require reductions in doses from all controlled sources combined to levels far below variations in natural background on the basis only of considerations of acceptable risk because, although the risk from natural background might not be welcome, the variations in background can hardly be called unacceptable (ICRP 1991). Furthermore, there is no direct evidence of increased risks due to radiation exposure at magnitudes of natural background (National Research Council 1990).

Second, levels of naturally occurring radionuclides in environmental media, including considerations of their bioavailability, can be important in developing standards for particular sources or practices. For example, cleanup standards for radium in soil at uranium and thorium mill-tailings sites were based in large part on background levels of radium in surface soil in parts of the United States where the mill tailings were produced. In this case, the maximum allowable concentrations of radium in soil must be substantially above background, so that the radium arising from mill tailings can be distinguished from the radium in native soil.

The influence of natural background radiation on the development of radiation standards is discussed further in chapter 7.

SUMMARY

This chapter has discussed the basic considerations that are important in developing any standards for controlling radiation exposures of the public. The main points of these discussions are summarized as follows.

First, all standards for radionuclides in the environment are based on judgments about the acceptability of magnitudes of health risk to the public or judgments about the achievability of magnitudes of health risk to the public. Those two considerations are applied without regard for any differences in mandates in the various environmental laws that provide the authority for regulations.

Second, other considerations also can be important in developing standards for radioactivity in the environment, including the justification of practices (positive net benefit), the ability to measure radioactivity in the environment at levels corresponding to the standards and the magnitude and variability of natural background radiation, and the levels of naturally occurring radionuclides in various environmental media. However, these other considerations usually do not provide the primary basis for standards for controlling radiation exposures of the public.

6
Organizations Concerned with Radiation Protection of the Public

INTRODUCTION

This chapter describes the important national and international organizations concerned with radiation protection of the public, including regulatory authorities in the United States and various national and international organizations that develop recommendations on radiation protection. In addition to a general discussion of the role of each organization in radiation protection of the public, the particular responsibilities for the development of standards for TENORM are emphasized. The standards for TENORM that have been developed by each organization are considered in more detail in chapters 7-10.

The principal federal agencies with responsibilities for radiation protection of the public are the Environmental Protection Agency (EPA), the Nuclear Regulatory Commission, and the Department of Energy (DOE). Of these, only EPA and DOE may develop guidances or regulations for TENORM. State governments also have important responsibilities for radiation protection of the public, including the development of regulations for TENORM. Finally, the National Council on Radiation Protection and Measurements (NCRP) and the Health Physics Society are important national organizations that have developed recommendations on radiation protection, including recommendations applicable to TENORM.

The International Commission on Radiological Protection (ICRP) is the principal international organization concerned with radiation protection. ICRP is an organization similar to NCRP and also develops recommendations on radiation protection. Other important international organizations are the International Atomic Energy Agency (IAEA) and the Commission of the European Communities (CEC).

ENVIRONMENTAL PROTECTION AGENCY

EPA was created by Reorganization Plan No. 3 of 1970. Its mission is to protect public health and to safeguard the natural environment (the air, water, and land), on which life depends. Under the authority of several laws, EPA develops environmental standards for both radiologic and nonradiologic hazards.

The responsibilities of EPA for radiation protection of the public are varied and complex. They include the development of federal guidance on radiation protection of the public; standards for radioactivity in the environment under authority of the Atomic Energy Act; standards for radioactivity under various laws, such as the Safe Drinking Water Act and Clean Air Act, that are concerned primarily with nonradiologic hazards; and guidance and regulations for indoor radon.

Federal Guidance on Radiation Protection of the Public

Executive Order 10831 assigned to EPA the responsibility for developing guidance for all federal agencies in the formulation of radiation-protection standards. This responsibility had been assigned previously to the Federal Radiation Council (FRC). The existing federal guidance on radiation protection of the public (FRC 1961; 1960) and EPA's proposed revision of the federal guidance (EPA 1994d) are discussed in chapter 7.

The federal guidance on radiation protection of the public presents basic, minimal requirements intended to ensure that reasonably consistent and adequately protective approaches are implemented by all federal agencies with regulatory responsibilities for radiation protection, especially the Nuclear Regulatory Commission and DOE. EPA is not authorized to enforce any provisions of the federal guidance, but all federal agencies are expected to comply with the guidance unless there are compelling reasons (such as specific statutory requirements) not to do so.

The federal guidance on radiation protection of the public is intended to apply to all controlled sources of exposure, including sources not associated with operations of the nuclear fuel cycle, but excluding indoor radon and beneficial medical exposures. Therefore, the federal guidance is intended to apply to all exposures of the public to TENORM, but not to naturally occurring radionuclides in their undisturbed state.

However, as indicated by the discussions in this chapter and in chapters 7 and 9, neither EPA nor any other federal agency with responsibilities for radiation protection of the public has developed standards that apply to all exposure situations involving TENORM. Rather, federal regulation of TENORM is rather fragmentary, and many potentially important sources of public exposure to TENORM are not regulated by any federal agency.

Environmental Radiation Standards Developed Under Atomic Energy Act

Under authority of the Atomic Energy Act, EPA is responsible for developing generally applicable environmental radiation standards for specific sources or practices associated with the nuclear fuel cycle that also are regulated by the Nuclear Regulatory Commission or DOE under the act (see later in this chapter). EPA has developed environmental standards under the Atomic Energy Act for operations of uranium fuel-cycle facilities, uranium and thorium mill tailings, and management and disposal of spent nuclear fuel, high-level waste, and transuranic waste; and it is considering the development of standards for management and disposal of low-level waste and cleanup of radioactively contaminated sites (see chapter 7).

Environmental radiation standards developed by EPA under the Atomic Energy Act do not apply to TENORM, because such materials are not defined in the act. Therefore, any environmental standard for TENORM developed by EPA must be authorized under some other law, as discussed in the next section.

An important feature of EPA's authority to establish environmental radiation standards for specific sources or practices under the Atomic Energy Act is that EPA usually is not the enforcement authority for these standards. Rather, they are enforced by the Nuclear Regulatory Commission or DOE in nearly all cases. This division of standard-setting and enforcement authorities between the EPA and the Nuclear Regulatory Commission or DOE is based on provisions of the Atomic Energy Act, which antedated the formation of EPA and assigned to the Atomic Energy Commission (AEC), a forerunner of the Nuclear Regulatory Commission, the responsibility for protecting public health and safety in the use of source, special nuclear, and byproduct materials arising from operations of the nuclear fuel cycle.

Environmental Radiation Standards Developed Under Other Laws

Under the authority of several laws other than the Atomic Energy Act, EPA is responsible for developing environmental standards for radionuclides and other nonradiologic hazards. In contrast with the standards developed by EPA under the Atomic Energy Act and discussed above, EPA also is the enforcement authority for its environmental standards developed under any other laws. The most important laws and their applicability to radionuclides associated with the nuclear fuel cycle and to TENORM are summarized as follows (the particular standards for radionuclides are discussed in more detail in chapter 7).

First, under authority of the Clean Air Act, EPA has established standards for airborne emissions of radionuclides from nuclear fuel-cycle facilities (that is, standards for airborne emissions of source, special nuclear, and byproduct materials which also are regulated under the Atomic Energy Act) and for particular airborne emissions of TENORM.

Second, under authority of the Clean Water Act, EPA may establish standards for release of naturally occurring and accelerator-produced radioactive materials (NARM), which include TENORM, to surface waters; such standards have been established for release of naturally occurring radionuclides from particular mines and mills. Source, special nuclear, and byproduct materials regulated under the Atomic Energy Act are excluded from regulation under the Clean Water Act, except that discharges of high-level waste into surface waters are banned.

Third, under authority of the Safe Drinking Water Act, EPA has established standards for naturally occurring and human-made radionuclides in public drinking-water supplies. For most public drinking-water supplies currently regulated under the Safe Drinking Water Act, exposures to naturally occurring radionuclides are the primary concern.

Fourth, under authority of the Comprehensive Environmental Response, Compensation, and Liability Act (CERCLA), which addresses environmental releases of hazardous substances that are not adequately regulated under other environmental laws, EPA has established regulations that define a process for selecting remedial actions, including the development of goals for cleanup of contaminated sites. The remediation process and goals for cleanup apply to sites contaminated with source, special nuclear, or byproduct materials or with NARM.

In addition, EPA may establish environmental standards for TENORM under the Toxic Substances Control Act (TSCA), which is concerned with protection of human health and the environment in the use of toxic substances in commerce, and the Resource Conservation and Recovery Act (RCRA), which is concerned in part with management and disposal of hazardous and nonhazardous solid waste. Particularly under TSCA, EPA could develop environmental standards for TENORM that apply to all aspects of management and disposal of these materials. However, EPA has not yet established standards for TENORM under either of these laws.

In summary, EPA is authorized to establish and enforce standards for radionuclides in the environment under several laws that are intended primarily to address nonradiologic hazards. In regard to TENORM, EPA has established standards for some airborne releases under the Clean Air Act, some releases to surface waters under the Clean Water Act, concentrations in public drinking-water supplies under the Safe Drinking Water Act, and cleanup of contaminated sites under CERCLA. EPA also may establish standards for TENORM under TSCA or RCRA, but it has not yet done so.

Thus, public exposures to TENORM are regulated by EPA only in a rather fragmentary manner under a variety of environmental laws, and no EPA regulation or set of regulations applies to all exposure situations involving TENORM. Such regulations could be developed under TSCA.

Guidance and Regulations for Indoor Radon

The Indoor Radon Abatement Act of 1988, which is Title III of TSCA, provides the authority for EPA's indoor-radon abatement program. The act requires EPA to issue guidance on mitigation of indoor radon (see chapters 7 and 8). This guidance is not a standard that is enforceable by EPA or the states, but it is widely used in the real estate and insurance industries. EPA also is authorized to issue regulations to carry out the provisions of the Indoor Radon Abatement Act. However, it has not yet issued any such regulations; furthermore, it is not authorized to establish enforceable standards for radon exposure or dose.

NUCLEAR REGULATORY COMMISSION

The Nuclear Regulatory Commission is an independent regulatory authority created by the Energy Reorganization Act of 1974, which amended the Atomic Energy Act by replacing AEC with the Nuclear Regulatory Commission and a separate agency called the Energy Research and Development Administration (ERDA), which later became part of DOE. The Energy Reorganization Act assigned to the Nuclear Regulatory Commission the previous responsibilities of AEC for protection of public health and safety in the use of source, special nuclear, and byproduct materials as defined in the Atomic Energy Act. Those responsibilities are carried out by means of the Nuclear Regulatory Commission's authority to license all commercial activities involving the use of these radioactive materials. The Nuclear Regulatory Commission also was given licensing authority over some facilities operated by DOE and other federal agencies.

In its role as an enforcement authority for environmental radiation standards established by EPA under the Atomic Energy Act, the Nuclear Regulatory Commission enforces any such standards that apply to licensed commercial activities, as well as the standards that apply to particular DOE activities.

The Nuclear Regulatory Commission derives all its regulatory authority from the Atomic Energy Act. Therefore, it has no regulatory authority over TENORM as defined in this study, because these materials do not arise from operations of the nuclear fuel cycle and are not defined in the act.

DEPARTMENT OF ENERGY

DOE was created by the Department of Energy Organization Act of 1977, which amended the Atomic Energy Act by combining ERDA, which had been created by the Energy Reorganization Act of 1974, with parts of other federal agencies. DOE is responsible for all atomic-energy defense activities and other activities involving energy research, development, and demonstration; it also is assigned the responsibility for protecting public health and safety in carrying out its authorized activities.

As noted above, DOE is an enforcement authority for environmental radiation standards established by EPA under the Atomic Energy Act. Specifically, DOE enforces any such standards that apply to activities of DOE or its contractors, with the exception of some activities for which EPA or the Nuclear Regulatory Commission is the enforcement authority.

As in the case of the Nuclear Regulatory Commission discussed in the previous section, the authority for all DOE activities derives from the Atomic Energy Act. However, unlike the Nuclear Regulatory Commission, DOE also has regulatory authority over any NARM and thus TENORM. That is its responsibility because DOE is required under the act to protect public health and safety in carrying out its authorized activities and EPA has not yet established environmental standards for NARM under TSCA. Current DOE requirements for management and disposal of TENORM contained in DOE Order 5400.5 (DOE 1990) are discussed in chapter 9.

STATE GOVERNMENTS

State governments have two important responsibilities for radiation protection of the public. First, the so-called agreement-state provisions of the Atomic Energy Act specify that the Nuclear Regulatory Commission may transfer to the states portions of its licensing authority over commercial uses of source, special nuclear, and byproduct materials. This state licensing authority does not apply to any other radioactive materials (such as TENORM) or to any DOE activities licensed by the Nuclear Regulatory Commission.

Second, in the absence of federal legislation and EPA regulations that address all exposure situations involving NARM and thus TENORM in the environment, these materials are subject to regulation only by the states, except for the fragmentary regulation of some sources and practices by EPA under a variety of environmental laws and the responsibilities of DOE for regulating these materials under its control. The authority for states to regulate NARM is based on the point of Constitutional law that any responsibilities for protection of public health and safety not specifically assigned to the federal government are delegated to the states.

Some states consider that TENORM is regulated under their general rules on radiation protection. However, as discussed in chapter 9, several states have developed regulations specifically for TENORM, and the Conference of Radiation Control Program Directors (CRCPD) has developed suggested state regulations for TENORM (CRCPD 1997) that are intended to provide guidance to the states in developing their own standards. The suggested state regulations have not been issued in final form, and they have not been implemented by any states.

NATIONAL COUNCIL ON RADIATION PROTECTION AND MEASUREMENTS

NCRP is a nonprofit corporation chartered by Congress in 1964 that, in addition to other authorized activities, develops recommendations on radiation protection. It is the successor organization to an unincorporated association of scientists called the National Committee on Radiation Protection and Measurements, which began as the Advisory Committee on X-ray and Radium Protection in 1929 and was reorganized as the National Committee on Radiation Protection in 1946.

NCRP is an advisory organization, and it has no authority to establish or enforce any requirements for radiation protection. However, its recommendations have been influential in the development of standards and guidances for radiation protection in the United States, initially by FRC, AEC, and the Public Health Service and later by the Nuclear Regulatory Commission, DOE, and EPA.

NCRP has issued many reports addressing NORM and TENORM (NCRP 1993b; 1993a; 1989a; 1988; 1987a; 1984c; 1984b). Current NCRP recommendations for control of exposures of the public to indoor radon and other NORM are discussed in chapters 8 and 9.

HEALTH PHYSICS SOCIETY

The Health Physics Society was formed in 1956 as a scientific organization concerned with protection of people and the environment from radiation. The society's membership includes professionals representative of all scientific and technical fields related to radiation protection and drawn from academe, government, medical institutions, research laboratories, and industry, both nationally and internationally. The society is chartered in the United States as a nonprofit scientific organization and is not affiliated with any government or industrial organization.

As a scientific organization, the Health Physics Society has no particular responsibilities for developing radiation-protection standards or recommendations. In recent years, however, the society's Scientific and Public Issues Committee has issued several position statements on policy matters of interest to radiation protection in response to opportunities to provide public comment. The position statement on standards for site cleanup and restoration and its relevance to regulation of TENORM are discussed in chapter 9.

INTERNATIONAL COMMISSION ON RADIOLOGICAL PROTECTION

ICRP was established in 1928 as the International X-ray and Radium Protection Committee and was restructured under its present name in 1950. It is an association of scientists from many countries, including the United States, that develops recommendations on all aspects of radiation protection. ICRP has official relationships with the World Health Organization and IAEA, and it has important relationships with other national and international organizations concerned with radiation protection, including NCRP.

Like NCRP, ICRP is an advisory organization with no authority to establish or enforce any requirements for radiation protection. However, its recommendations have greatly influenced the development of radiation-protection standards in many nations, including the United States.

ICRP has issued several reports addressing radon and occupational exposure to naturally occurring radionuclides in mines (ICRP 1993b; 1987b; 1986; 1981). Current ICRP recommendations for control of exposures of the public to indoor radon and other NORM are discussed in chapters 8 and 9.

INTERNATIONAL ATOMIC ENERGY AGENCY

IAEA is an intergovernmental organization established in 1957 under the auspices of the United Nations. It provides a forum for scientific and technical cooperation in nuclear activities, and it is the international inspectorate for the application of nuclear safeguards and verification measures covering nondefense nuclear programs.

One of IAEA's statutory objectives is to establish radiation-protection standards. Its current requirements for radiation protection, which essentially incorporate current ICRP recommendations (ICRP 1993b; 1991), are contained in the Basic Safety Standards (IAEA 1996a); these are intended to set forth requirements for member states, including the United States. However, the member states, not IAEA, are responsible for enforcing the standards. The Basic Safety Standards have not had any influence on the development of radiation standards in the United States beyond the influence of the ICRP recommendations that have provided the basis for the standards.

The Basic Safety Standards include provisions that apply to indoor radon and other naturally occurring radionuclides. These provisions are discussed in chapters 8 and 9.

COMMISSION OF THE EUROPEAN COMMUNITIES

CEC is authorized under the Euratom Treaty on Atomic Energy in the European Communities, which was signed in Rome in 1957, to establish radiation-protection standards for all members of the European Union. Enforcement of these standards is the responsibility of each member state, not CEC.

The current radiation-protection standards developed by CEC are based on current ICRP recommendations (ICRP 1991). However, as discussed in chapter 9, the CEC standards also include a special provision not included in the ICRP recommendations or in IAEA's Basic Safety Standards (IAEA 1996a) that addresses substantial increases in exposure due to natural radiation sources other than radon and would apply, for example, to operations, storage of materials, or residues containing naturally occurring radionuclides. Member states of the European Union are to develop requirements consistent with the provisions of CEC's radiation-protection standards by the year 2000 (Euratom 1996).

7
Environmental Protection Agency Guidances and Regulations for Naturally Occurring Radionuclides

The primary purpose of this chapter is to review existing and proposed Environmental Protection Agency (EPA) guidances and regulations that apply to control of routine exposures of the public to naturally occurring radionuclides. As discussed in chapter 2, the naturally occurring radionuclides of primary concern in radiation protection of the public include isotopes of uranium, thorium, and radium and their radiologically important shorter-lived decay products.

EPA guidances and regulations reviewed in this chapter include those that apply either to TENORM or to naturally occurring radionuclides associated with operations of the nuclear fuel cycle, which are not included in TENORM as defined in this study. No distinction is made in this review between TENORM and NORM associated with the nuclear fuel cycle because the intent is to indicate the variety of approaches used by EPA in regulating naturally occurring radionuclides for any exposure situations of concern. In chapter 10, EPA guidances and regulations that apply specifically to TENORM are summarized and compared with guidances for TENORM developed by other organizations.

The guidances and regulations considered in this review apply only to situations in which routine exposures to naturally occurring radionuclides are affected by human activities; they do not apply to naturally occurring radionuclides in their undisturbed state. This review is not concerned with EPA guidances on control of radiation exposures in the workplace (EPA 1987a) or responses to accidental releases of radionuclides to the environment (EPA 1992a).

EPA's guidances and regulations that apply to control of routine exposures of the public to naturally occurring radionuclides may be divided into two categories:

- Guidance on radiation protection of the public, which applies to all specified controlled sources of exposure combined.

- Guidance or regulations that apply only to specific sources or practices.

This review of EPA guidances and regulations emphasizes the quantitative criteria that apply to naturally occurring radionuclides and the basis for these criteria.

In addition to the specific guidances and regulations, this chapter discusses the health risks to the public that correspond to the quantitative criteria in different guidances and regulations, the important issue of consistency of standards in regard to limits on risk, and the relationship between the quantitative criteria in the various guidances and regulations and the doses or risks experienced in actual exposure situations.

GUIDANCE ON RADIATION PROTECTION OF THE PUBLIC

EPA is responsible for developing guidance for all federal agencies on standards for radiation protection of the public. These standards apply to all specified controlled sources of exposure combined, excluding indoor radon, but do not apply to natural background radiation and to beneficial medical exposures. EPA has issued proposed federal guidance on radiation protection of the public (EPA 1994d) to replace the guidance developed many years ago by the Federal Radiation Council (FRC 1961; 1960). Although the proposed guidance has not been issued in final form, the committee has assumed that it represents EPA's current views on the basic, minimal requirements for radiation protection of the public. Therefore, the proposed guidance is given greater emphasis in this study than the existing FRC guidance.

EPA's proposed federal guidance on radiation protection of the public includes the following provisions of interest to this study:

- There should be no radiation exposure of the general public unless it is justified by the expectation of an overall benefit from the activity causing the exposure.

- Doses to individuals and populations should be as low as reasonably achievable (ALARA).

- The annual effective dose equivalent to individuals from all controlled sources combined, including sources not associated with operations of the nuclear-fuel cycle but excluding indoor radon, should not normally exceed 1 mSv (100 mrem).

- Annual effective dose equivalents to individuals up to 5 mSv (500 mrem) may be permitted, with prior authorization, in unusual, temporary situations.

- Continued exposure over substantial portions of a lifetime at or near 1 mSv (100 mrem) per year should be avoided.

- Authorized limits for specific sources or practices should be established to ensure that the primary dose limit of 1 mSv (100 mrem) per year for all controlled sources combined and the ALARA objective are satisfied, and the authorized limit for any source or practice normally should be a fraction of the dose limit for all controlled sources combined.

The provisions listed above would apply to naturally occurring radionuclides, including TENORM, other than indoor radon, whenever exposures of the public are affected by human activities. However, to ensure compliance with these provisions, especially the primary dose limit for all sources of exposure combined, exposures to human-made radionuclides also would need to be taken into account.

EPA's proposed federal guidance was based in large part on recommendations on radiation protection of the public developed previously by the International Commission on Radiological Protection (ICRP 1977) and the National Council on Radiation Protection and Measurements (NCRP 1987c). In addition, the emphasis in the proposed guidance on the use of authorized limits for specific sources or practices at a fraction of the primary dose limit for all sources of exposure combined, to help ensure compliance with the primary dose limit and the ALARA objective, conforms to current recommendations of ICRP (1991) and NCRP (1993a).

The existing federal guidance on radiation protection of the public (FRC 1961; 1960), which EPA's proposed guidance would replace, includes the following provisions of interest to this study:

- There should not be any exposure to human-made radiation without the expectation of benefit from such exposure.

- Every effort should be made to encourage keeping radiation doses as far below recommended limits as practicable.

- For external exposure, the annual dose equivalent to the whole body of individuals should not exceed 5 mSv (500 mrem), and the dose equivalent to the gonads for average individuals in exposed populations should not exceed 50 mSv (5,000 mrem) in 30 y, that is, an average annual dose of 1.7 mSv (170 mrem).

- For internal exposure, the annual dose equivalent to individuals should not exceed 5 mSv (500 mrem) to bone marrow and 15 mSv (1,500 mrem) to bone or the thyroid, and the annual dose equivalents to these organs for average individuals in exposed populations should not exceed one-third of these values.

EPA's proposed federal guidance on radiation protection of the public differs from the existing FRC guidance in several important respects.

First, the proposed guidance is explicit that it would apply to all controlled sources of exposure combined (except as noted), including sources not associated with operations of the nuclear fuel cycle. The existing FRC guidance is not explicit on this point and has not been applied consistently to sources not associated with operations of the nuclear fuel cycle, especially to important sources of exposure to TENORM (EPA 1994d).

Second, the existing FRC guidance specifies dose limits for the whole body and the critical organ, and separate dose limits are specified for external and internal exposure. The proposed guidance would replace the dose limits for the whole body and the critical organ and the separate dose limits for external and internal exposure with a single limit on effective dose equivalent from external and internal exposure combined. The effective dose equivalent is intended to be proportional to stochastic risk posed by any exposure without regard for the distribution of doses among different organs or tissues.

Third, in most cases, the limit on annual effective dose equivalent of 1 mSv (100 mrem) in the proposed guidance is expected to correspond to lower allowable exposures than the existing FRC guidance on dose limits for the whole body or the critical organ from either external or internal exposure. The reduction in the maximum allowable exposures was based on information on the risk per unit dose that was not available when the FRC guidance was developed and on a judgment about an upper bound on acceptable risk posed by exposure to all controlled sources combined (see chapter 5).

Finally, the separate dose limit for the gonads of average individuals in the FRC guidance, which was intended to limit the induction of severe genetic effects in exposed populations, would no longer be specified. In the proposed guidance, the genetic risk would be taken into account in the weighting factor for the gonads used in defining the effective dose equivalent (ICRP 1977).

The essential purpose of EPA's proposed federal guidance on radiation protection of the public is to limit incremental health risks to exposed individuals and populations to levels that society generally regards as acceptable. In selecting the primary dose limit of 1 mSv (100 mrem) per year from exposure to all controlled sources combined, EPA considered several judgmental factors. These factors included the lifetime risk corresponding to the limit on annual dose, the degree of additional protection that would be achieved by the application of ALARA by regulatory authorities for specific sources or practices and by the consideration of the possibilities for multiple exposures to current and future sources, and the record on the operational application of the ALARA objective in reducing actual doses to levels below authorized limits (EPA 1994d).

The lifetime risk corresponding to the primary dose limit of 1 mSv (100 mrem) per year can be estimated by assuming continuous exposure over 70 y at the dose limit and a risk of fatal cancers per unit dose for members of the public of 5×10^{-5} per millisievert (5×10^{-7} per millirem) (EPA 1994c; NCRP 1993a; ICRP 1991). On the basis of those assumptions, the lifetime risk of fatal cancers would be about 4×10^{-3}. This value is somewhat higher than the lifetime risk of about 10^{-3} that ICRP (1977) judged to be an upper bound on acceptable risk posed by radiation exposure on the basis of data on other involuntary risks that the public has accepted in everyday life (see chapter 5).

However, as emphasized in the proposed federal guidance (EPA 1994d), compliance with the primary dose limit of 1 mSv (100 mrem) per year does not, by itself, provide acceptable radiation protection of the public; compliance with the ALARA objective also is a central tenet of radiation protection. Indeed, as a result of the establishment of authorized limits for specific sources or practices at a fraction of the primary dose limit and further vigorous application of the ALARA objective at specific sites, the average annual effective dose equivalent to individuals in exposed populations within 80 km (50 miles) of operating nuclear facilities is only about 0.5 µSv (0.05 mrem) (NCRP 1987a). That dose corresponds to a lifetime risk of fatal cancers of only about 2×10^{-6}, or lower by a factor of 2,000 than the risk corresponding to the primary dose limit. Furthermore, doses to individuals receiving the highest exposures, although they might substantially exceed the average dose in exposed populations, normally are only about 10% of the primary dose limit or less (EPA 1989d).

Thus, although the purpose of the proposed federal guidance is to limit risks posed by radiation exposure, an acceptable risk is not defined by the primary dose limit alone. For most exposure situations, the acceptability of risks is defined primarily by application of the ALARA objective, which involves judgments about doses to individuals and populations that are reasonably achievable for specific sources or practices and at specific sites. Even though compliance with the ALARA objective can be defined to some extent by authorized limits for specific sources or practices at a fraction of the primary dose limit, application of the objective at each site is a process, not a result that can be specified a priori in regulations.

An additional factor taken into account by EPA in judging that the primary dose limit of 1 mSv (100 mrem) per year and reductions in dose below the limit to meet the ALARA objective would provide acceptable risks to individuals and populations was the unavoidable risk posed by exposure to natural background radiation. The average effective dose equivalent from all natural sources—including cosmic rays, cosmogenic and terrestrial radionuclides, radionuclides in the body, and indoor radon—is about 3 mSv (300 mrem) per year in the United States (see table 2.10). The primary dose limit proposed by EPA thus corresponds to about one-third of the average dose from natural background radiation, for which the estimated lifetime risk of fatal cancers is about 10^{-2}. Although the average dose from exposure to natural background does not provide a justification for the primary dose limit for all controlled sources combined, it does provide a perspective for judging whether the dose limit for all controlled sources is reasonable (see chapter 5).

GUIDANCE AND REGULATIONS FOR SPECIFIC SOURCES OR PRACTICES

EPA is authorized under several environmental laws to establish guidance or regulations for controlling radiation exposures of the public to specific sources or practices (see chapter 6). As noted in the previous section, authorized limits for specific sources or practices (also called source constraints or dose constraints) are an important means of ensuring compliance with the primary dose limit for all controlled sources combined and the ALARA objective in radiation-protection standards for the public.

EPA's guidances and regulations for specific sources or practices can be divided into the following categories (the legal authority for establishing guidance or regulations in each category is given in parentheses):

- Operations of uranium fuel-cycle facilities (Atomic Energy Act).
- Radioactivity in drinking water (Safe Drinking Water Act).
- Radioactivity in liquid discharges (Clean Water Act).

- Uranium and thorium mill tailings (Uranium Mill Tailings Radiation Control Act; Atomic Energy Act).

- Radioactive waste management and disposal (Atomic Energy Act).

- Remediation of radioactively contaminated sites (Comprehensive Environmental Response, Compensation, and Liability Act, CERCLA; Atomic Energy Act).

- Airborne emissions of radionuclides (Clean Air Act).

- Indoor radon (Indoor Radon Abatement Act).

In addition, EPA may, under the Toxic Substances Control Act (TSCA), regulate naturally occurring and accelerator-produced radioactive materials (NARM), including TENORM, which are not subject to regulation under the Atomic Energy Act; and NARM wastes also could be regulated under the Resource Conservation and Recovery Act (RCRA). EPA has not developed proposed regulations specifically for NARM under either TSCA or RCRA.

The following sections review existing or proposed guidances and regulations for specific sources or practices developed by EPA that apply to naturally occurring radionuclides. The relevant quantitative criteria in the guidances and regulations are presented, and the bases for the criteria are discussed. The criteria that apply to naturally occurring radionuclides, including EPA's proposed federal guidance on radiation protection of the public discussed above, also are summarized in table 7.1. After the discussions of the guidances and regulations, the possibility of regulating NARM under TSCA or RCRA is discussed.

Table 7.1. Summary of EPA guidances and regulations applicable to naturally occurring radionuclides[a]

Guidance or regulation	Quantitative criteria[b]	Comments
Proposed federal guidance on radiation protection of the public (EPA 1994d)[c]	Annual effective dose equivalent of 1 mSv	Dose limit applies to all controlled sources of exposure combined, excluding indoor radon and beneficial medical exposures. Based on considerations of maximum tolerable risk to individuals and ability of authorized limits for specific sources or practices and further application of ALARA objective to reduce doses well below limit.
Standards for operations of uranium fuel-cycle facilities (40 CFR Part 190)	Annual dose equivalent of 0.25 mSv to whole body, 0.75 mSv to thyroid, and 0.25 mSv to any other organ	Based primarily on doses judged reasonably achievable with available effluent-control technologies.
Interim standards for radioactivity in community drinking-water systems (40 CFR Part 141)	Concentration of 5 pCi/L for ^{226}Ra plus ^{228}Ra[d] Concentration of 15 pCi/L for gross alpha-particle activity, including ^{226}Ra but excluding radon and uranium[d]	Based primarily on cost-benefit analysis for reducing existing levels of naturally occurring radionuclides in drinking water.
Proposed revisions of interim standards for radioactivity in community drinking-water systems (EPA 1997; 1991a)	Concentration of 20 pCi/L for ^{226}Ra and ^{228}Ra separately[d] Concentration of 20 µg/L for uranium Concentration of 15 pCi/L for gross alpha-particle activity, excluding ^{226}Ra, uranium, and ^{222}Rn[d] Annual effective dose equivalent of 0.04 mSv from all beta- or gamma-emitting radionuclides, excluding ^{228}Ra	Based primarily on revised cost-benefit analysis for reducing existing levels of naturally occurring radionuclides in drinking water.

Table 7.1. (continued)

Guidance or regulation	Quantitative criteria	Comments
Standards for radioactivity in liquid discharges (40 CFR Part 440)	Concentrations in daily effluents of 10 pCi/L for dissolved ^{226}Ra, 30 pCi/L for total ^{226}Ra, and 4 mg/L for uranium[d] Average concentrations in daily effluents over 30 d of 3 pCi/L for dissolved ^{226}Ra, 10 pCi/L for total ^{226}Ra, and 2 mg/L for uranium[d]	Limits apply to liquid discharges from mines or mills used to produce or process uranium, radium, or vanadium ores. Based primarily on available effluent-control technologies.
Standards for uranium or thorium mill tailings (40 CFR Part 192)	Annual average release rate of ^{222}Rn to air of 20 pCi/m^2 per second or concentration of ^{222}Rn in air outside disposal site of 0.5 pCi/L[d] Average concentrations of ^{226}Ra in soil above background over any area of 100 m^2 of 5 pCi/g in top 15 cm or 15 pCi/g below 15 cm[d] Concentration of radon decay products indoors including background of 0.03 WL, with objective of 0.02 WL[e] Indoor gamma radiation level above background of 20 µR/h[f] Concentrations in groundwater of 5 pCi/L for ^{226}Ra plus ^{228}Ra, 15 pCi/L for gross alpha-particle activity, and 30 pCi/L for ^{234}U plus ^{238}U[d] Annual dose equivalent from thorium-processing operations as in uranium fuel-cycle standards (40 CFR Part 190)	Releases during uranium-processing operations and from uranium mill tailings disposal sites before end of closure period must comply with dose constraint in 40 CFR Part 190 and concentration limits for liquid discharges in 40 CFR Part 440. Based primarily on background levels of radioactivity in western United States and objective of reducing exposures of the public to as close to background levels as reasonably achievable; groundwater-protection requirements are based on current and proposed drinking-water standards (40 CFR Part 141).

Table 7.1. (continued)

Guidance or regulation	Quantitative criteria	Comments
Standards for management and storage of spent fuel, high-level waste, and transuranic waste (40 CFR Part 191)	For facilities regulated by Nuclear Regulatory Commission or Agreement States, annual dose equivalent of 0.25 mSv to whole body, 0.75 mSv to thyroid, and 0.25 mSv to any other organ For DOE facilities not regulated by Nuclear Regulatory Commission or Agreement States, annual dose equivalent of 0.25 mSv to whole body and 0.75 mSv to any organ	Based primarily on doses judged reasonably achievable with available effluent-control technologies; dose constraint is consistent with uranium fuel-cycle standards (40 CFR Part 190).
Standards for disposal of spent fuel, high-level waste, and transuranic waste (40 CFR Part 191)[g]	Cumulative releases to accessible environment per 1,000 MTHM of 100 Ci for ^{226}Ra, ^{234}U, ^{235}U, and ^{238}U; 10 Ci for ^{230}Th and ^{232}Th; and 1,000 Ci for ^{210}Pb[h] Annual effective dose equivalent in accessible environment from all exposure pathways of 0.15 mSv[i] Levels of radioactivity in underground sources of drinking water in accessible environment as specified by MCLs in drinking-water standards (40 CFR Part 141)[j]	Cumulative release limits were based on 1,000 health effects in US population, which was judged reasonably achievable. Dose constraint for individuals was based on judgment about acceptability of risk and feasibility of achieving specified dose. Groundwater protection requirement was based on general strategy of protecting resource consistent with current drinking-water standards.
Standards for cleanup of radioactively contaminated sites (CERCLA and 40 CFR Part 300)	Goal of compliance with ARARs, TBCs, and lifetime cancer risk of 10^{-4}; limits that must be achieved by cleanups without regard for other factors are not specified	Based on goal of complying with relevant requirements under other environmental laws and achieving consistency with cancer risks corresponding to other laws and regulations (such as Safe Drinking Water Act and Clean Air Act).

Table 7.1. (continued)

Guidance or regulation	Quantitative criteria	Comments
Standards for airborne emissions of radionuclides (40 CFR Part 61)	Annual effective dose equivalent of 0.1 mSv for many DOE and non-DOE federal facilities, but excluding dose from ^{222}Rn and its decay products, and for emissions of ^{222}Rn from underground uranium mines Annual emissions of ^{210}Po from elemental phosphorus plants of 2 or 4.5 Ci[h] Emission rate of ^{222}Rn from specified radium-bearing materials of 20 pCi/m^2 per second[d]	Based primarily on lifetime cancer risk to maximally exposed individuals of 10^{-4} and average lifetime risk in exposed populations of 10^{-6}.
Guidance on radon in homes (EPA and DHHS 1994)	Mitigation for radon concentrations above 4 pCi/L[d] Mitigation for radon concentrations of 2-4 pCi/L if concentrations can be reduced below 2 pCi/L[d]	Based on protection of individuals receiving highest exposures and cost-benefit analysis for reducing existing levels of radon in homes.

[a]Guidances or regulations that do not specifically apply only to naturally occurring radionuclides apply to human-made and naturally occurring radionuclides combined. Guidances or regulations that apply only to human-made radionuclides are not given in the table.
[b]Criteria expressed in terms of dose equivalent apply to individual members of the public. Criteria expressed in terms of quantities other than dose are given in the units presented in the guidance or regulation, and the conversion to SI units is indicated in a footnote.
[c]Proposed guidance would replace existing guidance of Federal Radiation Council (FRC 1961; 1960), which essentially specifies limit on annual dose equivalent of 5 mSv.
[d]1 pCi = 0.037 Bq.
[e]1 Working Level (WL) = 2.08 x 10^{-5} J/m^3.
[f]1 R = 2.58 x 10^{-4} C/kg.
[g]Standards apply for 10,000 years after disposal.
[h]1 Ci = 0.037 TBq.
[i]Standard applies only to undisturbed performance of disposal system (that is, absent human intrusion).

Standards for Operations of Uranium Fuel Cycle

EPA's current standards for operations of uranium fuel-cycle facilities in 40 CFR Part 190 were established in 1977 (42 FR 2858).[4] They apply to normal operations in the milling of uranium ore, chemical conversion of uranium, isotopic enrichment of uranium, fabrication of uranium fuel, electricity generation in light-water-cooled nuclear power plants using uranium fuel, and reprocessing of spent uranium fuel, but not to mining operations, transportation of radioactive material, operations at waste-disposal sites, and reuse of recovered non-uranium special nuclear material and byproduct materials as defined in the Atomic Energy Act.

The particular standard that applies to releases of naturally occurring radionuclides is a constraint on annual dose equivalent to individuals from all radionuclides, except radon and its decay products, of:

- 0.25 mSv (25 mrem) to the whole body.
- 0.75 mSv (75 mrem) to the thyroid.
- 0.25 mSv (25 mrem) to any other organ.

Separate activity limits on releases of some longer-lived, human-made radionuclides also are specified, but such limits are not specified for any naturally occurring radionuclides.

The dose constraint in the uranium fuel-cycle standards given above is essentially 5% of the primary whole-body dose limit of 5 mSv (500 mrem) per year for exposure to all controlled sources combined in the existing FRC guidance discussed earlier. However, the dose constraint in the fuel-cycle standards was based not on a judgment that doses at these levels were necessary to achieve acceptable health risks to the public, but primarily on a judgment that the specified doses were reasonably achievable with available effluent-control technologies. Thus, the standard essentially represents an application of the ALARA objective to standard-setting itself. An additional factor in establishing

---

[4]Title 40 of the *Code of Federal Regulations* (CFR), published annually by the US Government Printing Office, contains all EPA regulations. For each regulation published in the CFR, reference to the *Federal Register* (FR) notice containing the promulgated regulation is given; these notices provide supplementary information on the basis for the regulations.

the dose constraint was that the corresponding levels of radioactivity in the environment should be readily measurable (see chapter 5).

In the time since EPA's uranium fuel-cycle standards were promulgated, an authorized limit of 0.25 mSv (25 mrem) per year has been incorporated in other EPA standards for specific sources or practices, as well as in standards for low-level waste disposal established by the Nuclear Regulatory Commission (1982a) and the Department of Energy (DOE 1988). Furthermore, on the basis of the currently accepted risk per unit dose of 5×10^{-5} per millisievert and an assumption that the lifetime risk posed by exposure to all controlled sources combined should not exceed about 10^{-3}, an authorized limit of 0.25 mSv (25 mrem) per year for specific human-made sources is now widely regarded as necessary for protection of public health (for example, NCRP 1993a). Thus, a dose constraint of 0.25 mSv (25 mrem) per year for specific sources or practices has attained an importance for radiation protection of the public considerably beyond its original use in the uranium fuel-cycle standards.

Standards for Radioactivity in Drinking Water

EPA's current (interim) standards for radioactivity in community drinking-water systems in 40 CFR Part 141 were established in 1976 (41 FR 28404). They are concerned primarily with exposures to naturally occurring radionuclides, principally radium, and they apply at the tap rather than at the source.

The current drinking-water standards that apply to naturally occurring radionuclides include concentration limits of:

- 0.2 Bq/L (5 pCi/L) for radium-226 plus radium-228.

- 0.6 Bq/L (15 pCi/L) for gross alpha-particle activity, including ^{226}Ra but excluding radon and uranium.

The standards for naturally occurring radionuclides are expressed in terms of concentration rather than dose to individuals to allow compliance to be monitored by operators of water systems.

The interim standards also include a dose constraint of 0.04 mSv (4 mrem) per year to the whole body or any organ for human-made, beta- or gamma-emitting radionuclides, but this standard does not apply to any naturally occurring radionuclides.

The drinking-water standards for radionuclides were developed in accordance with requirements of the Safe Drinking Water Act. The act requires, first, that EPA establish maximum contaminant level goals (MCLGs), which are nonenforceable health goals that must be set at levels where no known or anticipated health effects would occur and an adequate margin of safety for protecting public health is provided. For known carcinogens, including radionuclides, the MCLGs must be set at zero on the basis of the usual assumption, for purposes of risk management, of a linear, no-threshold dose-response relationship. The act then requires that EPA establish maximum contaminant-levels (MCLs) for drinking water. The MCLs are legally enforceable standards that must be set as close to the MCLGs as possible with technical feasibility and cost taken into account.

Therefore, particularly for radium, which was regarded as the most important naturally occurring radionuclide in drinking water, EPA developed the drinking-water standard primarily on the basis of an analysis of the costs of reducing radioactivity in drinking water in relation to the benefits in health risks averted. This approach essentially represents another application of the ALARA objective to standard-setting itself. Although a risk assessment was performed in developing the standards, an a priori judgment about an acceptable risk posed by radionuclides in drinking water was not a consideration in establishing the standards.

The dose corresponding to the standard for radium in drinking water can be estimated by assuming that the water contains ^{226}Ra at 0.2 Bq/L (5 pCi/L) and that an individual ingests 2 L of water per day (EPA 1989c). For the effective dose equivalent per unit activity intake of ^{226}Ra given in current federal guidance (Eckerman and others 1988), the estimated effective dose equivalent is about 0.05 mSv (5 mrem) per year. The dose corresponding to the standard for ^{228}Ra in drinking water is nearly the same.

In 1991, EPA issued proposed revisions of the interim standards for radioactivity in community drinking-water systems (EPA 1991a). The proposed standards include the following criteria that apply to naturally occurring radionuclides:

- A concentration limit of 0.7 Bq/L (20 pCi/L) for ^{226}Ra and ^{228}Ra separately.

- A concentration limit of 20 µg/L for uranium.

- A concentration limit of 11 Bq/L (300 pCi/L) for radon-222.

- A concentration limit of 0.6 Bq/L (15 pCi/L) for gross alpha-particle activity, excluding ^{226}Ra, uranium, and ^{222}Rn.

- A limit on annual effective dose equivalent of 0.04 mSv (4 mrem) for all beta- or gamma-emitting radionuclides, excluding ^{228}Ra.

The proposed standard for uranium was based on prevention of chemical toxicity in the kidney, as well as considerations of cancer risk posed by radiation exposure. On the basis of the observed activity disequilibrium between uranium-238 and its decay product uranium-234 in natural waters and the recognition that the activity of uranium-235 in water is insignificant, the proposed limit on mass concentration for uranium was assumed to correspond to an activity concentration of 1.1 Bq/L (30 pCi/L) (EPA 1991a).

The proposed revisions of the drinking-water standards differ from the current standards in the following respects. First, the concentration limit for ^{226}Ra plus ^{228}Ra would be increased by a factor of 8, on the basis of a revised cost-benefit analysis (EPA 1991a). Second, uranium and ^{222}Rn in drinking water would be regulated for the first time. Third, the standard for gross alpha-particle activity would exclude ^{226}Ra. Finally, the standard for beta- or gamma-emitting radionuclides, which currently applies only to human-made radionuclides, would also apply to naturally occurring radionuclides other than ^{228}Ra (for example, to lead-210) and would be expressed in terms of the effective dose equivalent rather than the dose to the whole body or any organ.

The Safe Drinking Water Act Amendments of 1996 contain two provisions that directly affect the proposed revisions of the drinking-water standards for radionuclides. First, in response to the controversy over the cost-benefit analysis of the proposed standard for radon, the amendments directed that this proposal be withdrawn and that a new standard for radon in drinking water be promulgated by the year 2000 on the basis of results of a study by the National Academy of Sciences; in response to this directive, EPA has withdrawn the proposed standard for radon (EPA 1997).

Second, the amendments specify that any revision of drinking-water standards shall maintain or increase health protection of the public. The proposal to increase the allowable concentrations for radium clearly would result in higher allowable risks. The proposed revision of the standard for beta- or gamma-emitting radionuclides also would result in higher allowable risks because the proposed limit on the effective dose equivalent of 0.04 mSv (4 mrem) per year generally results in higher allowable concentrations of radionuclides in drinking water than the same limit on the dose equivalent to any organ or tissue (Eckerman and others 1988). Therefore, promulgation of the proposed revisions of drinking-water standards for radium and beta- or gamma-emitting radionuclides appears to be precluded by the amendments.

Standards for Radioactivity in Liquid Discharges

Under authority of the Clean Water Act, EPA develops standards aimed at restoring and maintaining the chemical, physical, and biologic integrity of the nation's surface waters. In particular, EPA may establish standards for release of naturally occurring radionuclides into surface waters (see chapter 6).

In 1982, EPA established standards in 40 CFR Part 440 for liquid discharges of naturally occurring radionuclides from mines or mills used to produce or process uranium, radium, and vanadium ores (47 FR 54609). These standards include the following provisions:

- Limits on concentrations in effluents for any day of 0.4 Bq/L (10 pCi/L) for dissolved ^{226}Ra, 1.1 Bq/L (30 pCi/L) for total ^{226}Ra, and 4 mg/L for uranium.

- Limits on average concentrations in daily effluents for 30 consecutive days of 0.11 Bq/L (3 pCi/L) for dissolved ^{226}Ra, 0.4 Bq/L (10 pCi/L) for total ^{226}Ra, and 2 mg/L for uranium.

Those limits were based on considerations of the effectiveness of effluent control technologies, rather than potential health risks to the public posed by ingestion of contaminated surface water.

EPA has not developed standards under the Clean Water Act for discharges of naturally occurring radionuclides to surface waters from any other sources. However, as noted in the following section, the standards in 40 CFR Part 440 apply to discharges from active uranium- and thorium-processing sites associated with the nuclear fuel cycle.

Standards for Uranium and Thorium Mill Tailings

EPA's current standards for uranium and thorium mill tailings in 40 CFR Part 192 were first established in 1983 (48 FR 602 and 48 FR 45946), and the provisions for groundwater protection were revised in 1995 (60 FR 2854). Those standards are concerned with control and cleanup of residual radioactive materials at or near inactive uranium- and thorium-processing sites and management of uranium and thorium byproduct materials at active processing sites. Only naturally occurring radionuclides are found in mill tailings, and the most important radionuclides of concern in protecting public health are radium, radon, and their decay products.

The mill-tailings standards are contained in four Subparts that apply to different aspects of management, disposal, or remediation, as follows.

Subpart A For control of residual radioactive materials from inactive uranium-processing sites, that is, uranium mill-tailings piles at inactive processing sites managed by DOE:

- A limit on annual average release rate to the atmosphere of 0.7 Bq/m² (20 pCi/m²) per second for ^{222}Rn or a limit on annual average concentration of 20 Bq/m³ (0.5 pCi/L) for ^{222}Rn in air above background outside the disposal site.

- Limits on concentrations in groundwater of 0.2 Bq/L (5 pCi/L) for ^{226}Ra plus ^{228}Ra, 0.6 Bq/L (15 pCi/L) for gross alpha-particle activity excluding radon and uranium, and 1.1 Bq/L (30 pCi/L) for ^{234}U plus ^{238}U, with provisions for establishing alternative concentration limits.

- Controls for limiting radon emissions and releases to groundwater designed to be effective for up to 1,000 y to the extent reasonably achievable and, in any case, for at least 200 y.

Subpart B For cleanup of land and buildings contaminated with residual radioactive materials from inactive uranium-processing sites, that is, for contaminated land and buildings at processing sites managed by DOE and contaminated real property in the vicinity of such sites:

- In land averaged over any area of 100 m², limits on concentrations of ^{226}Ra in soil above background of 0.2 Bq/g (5 pCi/g) averaged over the first 15 cm below the ground surface and 0.6 Bq/g (15 pCi/g) averaged over any 15-cm-thick layers more than 15 cm below the ground surface.

- In any occupied or habitable building, a limit on concentration of radon decay products, including background, of 6 x 10^{-7} J/m³ (0.03 working level (WL)), with an objective for remedial action of 4 x 10^{-7} J/m³ (0.02 WL).

- In any occupied or habitable building, a limit on gamma radiation level above background of 20 µR/h.

- Compliance with the groundwater protection standard in Subpart A.

Subpart D For management of uranium byproduct materials at active uranium processing sites, that is, for uranium-processing sites licensed by the Nuclear Regulatory Commission or Agreement States (states that enter into licensing agreements with the Nuclear Regulatory Commission):

- During processing operations and before the end of the closure period, compliance with the groundwater-protection standard in Subpart A, except for the concentration limit for ^{234}U plus ^{238}U; the flux standard for ^{222}Rn from tailings piles in Subpart A, but not the concentration standard outside the site; the dose constraint for individual members of the public in 40 CFR Part 190 (see above); and the limits on radioactivity in liquid discharges to surface waters in 40 CFR Part 440 (see above).

- After the closure period and for the period specified in Subpart A, compliance with the flux standard for ^{222}Rn from tailings piles in Subpart A, but not the concentration standard outside the site, except that the flux standard does not apply to any portion of a site that contains concentrations of ^{226}Ra in soil above background less than the values specified in Subpart B.

Subpart E For management of thorium byproduct materials at active thorium-processing sites, that is, for thorium-processing sites licensed by the Nuclear Regulatory Commission or Agreement States:

- Application of the standards for uranium, ^{222}Rn, and ^{226}Ra in Subpart D to thorium, ^{220}Rn, and ^{228}Ra, respectively, except that the flux standard for ^{222}Rn during uranium-processing operations and before the end of the closure period does not apply to releases of ^{220}Rn at thorium-processing sites during the same period.

- During thorium-processing operations and before the end of the closure period, limits on annual dose equivalent to individual members of the public of 0.25 mSv (25 mrem) to the whole body, 0.75 mSv (75 mrem) to the thyroid, and 0.25 mSv (25 mrem) to any other organ, excluding the dose from ^{220}Rn and its short-lived decay products.

Except for the dose constraint for individual members of the public in Subparts D and E that applies during uranium- or thorium-processing operations and before the end of the closure period, the mill-tailings standards are expressed in terms of quantities that can be measured in the field, rather than dose to individuals.

The standards for cleanup of residual radioactive materials in Subpart B, except for the groundwater-protection standards, are interrelated. Specifically, the concentration limit for ^{226}Ra in surface soil of 0.2 Bq/g

(5 pCi/g) is intended to ensure that the concentration of indoor radon decay products would be less than the objective of 4×10^{-7} J/m^3 (0.02 WL) and that the indoor gamma radiation level above background would be less than 20 μR/h (EPA 1982). However, the concentration limit for ^{226}Ra in subsurface soil of 0.6 Bq/g (15 pCi/g) is not health-based but is intended only to provide a standard that would allow the detection of mill tailings in subsurface soil by direct gamma measurement in the field (EPA 1982). As discussed in chapter 9, the two cleanup criteria for ^{226}Ra in soil in Subpart B often have been used in standards for TENORM.

Most of the provisions in the mill-tailings standards, especially the standards for cleanup of contaminated land and buildings, were based primarily on background levels of radioactivity in areas of the western United States where uranium and thorium ore deposits exist and the residual radioactive materials were obtained (EPA 1982). In addition, the standards for groundwater protection were based on the interim drinking-water standards in 40 CFR Part 141 with an additional provision for uranium, and the standards for thorium-processing operations and before the end of the closure period were based on the uranium fuel-cycle standards in 40 CFR Part 190.

The standard for indoor radon decay products also has been discussed by Harley (1996). If the activity of the decay products in indoor air is assumed to be 50% of the activity of radon, then the objective for remedial action of 4×10^{-7} J/m^3 (0.02 WL) corresponds to a radon concentration of about 150 Bq/m^3 (4 pCi/L), which is the current EPA guideline for mitigation of indoor radon discussed later in this chapter and in chapter 8.

As described below, the standards for inactive uranium-mill tailings sites can be converted to estimates of dose to individuals residing on contaminated land near the site. The most important contributors to dose are radium in soil, outdoor and indoor radon, and indoor gamma radiation.

An upper bound on the external dose corresponding to the concentration limits for ^{226}Ra in soil can be estimated by assuming continuous external exposure; the presence of all decay products of ^{226}Ra in equilibrium; indoor and outdoor residence times of 85% and 15%, respectively; a dose-reduction factor during indoor residence due to building shielding of 0.7 (Nuclear Regulatory Commission 1977); and external dose rates per unit concentration of ^{226}Ra in surface soil as given in current federal guidance (Eckerman and Ryman 1993). On the basis of those assumptions, the estimated annual effective dose equivalent from external exposure is about 0.5 mSv (50 mrem). For mill tailings, the dose from internal exposure to radium and its decay products, except for inhalation of radon decay products (which is considered separately), is expected to be considerably less than the dose from external exposure (EPA 1982).

The dose corresponding to the standard for outdoor radon of 20 Bq/m^3 (0.5 pCi/L) can be estimated by assuming the mean annual effective dose equivalent per unit concentration for an outdoor residence time of 15% recommended by (ICRP 1987b). On the basis of that assumption, the estimated annual effective dose equivalent is about 0.3 mSv (30 mrem).

The dose corresponding to the standard for indoor radon decay products of 6 x 10^{-7} J/m^3 (0.03 WL) can be estimated by assuming the mean annual effective dose equivalent per unit exposure for an indoor residence time of 85% recommended by ICRP (1987b). On the basis of that assumption, the estimated annual effective dose equivalent is about 8 mSv (800 mrem).

Finally, the dose corresponding to the standard for indoor gamma radiation can be estimated by assuming that an exposure of 1 R corresponds to an effective dose equivalent of about 7 mSv (700 mrem) (ICRP 1987a) and an indoor residence time of 85%. On the basis of those assumptions, the estimated annual effective dose equivalent is about 1 mSv (100 mrem). That includes the contribution from external exposure to radium in soil.

On the basis of those calculations, the mill-tailings standards correspond to a maximum annual effective dose equivalent to individual members of the public of nearly 10 mSv (1,000 mrem), and the contribution from all sources of exposure other than radon is about 1 mSv (100 mrem). The dose from all sources other than radon is less by about a factor of 5 than the primary dose limit of 5 mSv (500 mrem) per year from all controlled sources combined in the existing FRC guidance on radiation protection of the public (FRC 1960) but is essentially the same as the primary dose limit of 1 mSv (100 mrem) from all controlled sources combined in EPA's proposed revision of the federal guidance (EPA 1994d). Therefore, compliance with the recommendation in the proposed revision of the federal guidance that the dose from individual sources or practices normally should be limited to a fraction of the dose limit of 1 mSv (100 mrem) per year appears to be impractical for properties in the vicinity of uranium mill tailings disposal sites.

The dose calculations described above apply only to individuals residing on contaminated properties near uranium mill tailings sites, but they do not apply to individuals who might reside on a tailings pile itself at some time in the future. Given that undiluted mill tailings typically contain ^{226}Ra at about 10-40 Bq/g (300-1,000 pCi/g) (DOE 1996), in contrast with the cleanup standard for ^{226}Ra in surface soil of 0.2 Bq/g (5 pCi/g), permanent residence on an exposed tailings pile (for example, a pile whose cover had been removed inadvertently) would result in doses from external exposure and exposure to indoor radon that are about 2 orders of magnitude higher than the dose estimates for residence on contaminated land near the site. Such high doses clearly would be unacceptable under any circumstances. However, the intent under the Uranium Mill Tailings Radiation Control Act is that the high doses that could result from residence on a mill-tailings pile would be prevented by maintaining perpetual federal control over the sites to preclude permanent occupancy by members of the public.

Standards for Management and Disposal of Radioactive Waste

Under authority of the Atomic Energy Act, EPA has established standards for management and disposal of spent nuclear fuel, high-level waste, and transuranic waste. EPA also is developing standards for low-level waste.

Standards for Spent Fuel, High-Level Waste, and Transuranic Waste EPA's current standards for spent nuclear fuel, high-level waste, and transuranic waste in 40 CFR Part 191 were first established in 1985 (50 FR 38066) and then revised in 1993 (58 FR 66398). The standards apply to management (except for transportation), storage, and disposal of waste, and they apply to naturally occurring radionuclides in the wastes.

The standards for management and storage of spent fuel, high-level waste, and transuranic waste include the following provisions:

- For facilities regulated by the Nuclear Regulatory Commission or Agreement States, and including all operations of uranium fuel-cycle facilities covered by 40 CFR Part 190, limits on annual dose equivalent to individuals of 0.25 mSv (25 mrem) to the whole body, 0.75 mSv (75 mrem) to the thyroid, and 0.25 mSv (25 mrem) to any other organ.

- For facilities operated by DOE and not regulated by the Nuclear Regulatory Commission or Agreement States, limits on annual dose equivalent to individuals of 0.25 mSv (25 mrem) to the whole body and 0.75 mSv (75 mrem) to any organ or, on application for an alternative standard, limits on annual dose equivalent from all sources combined, excluding natural background and medical practices, of 1 mSv (100 mrem) for continuous exposure or 5 mSv (500 mrem) for infrequent exposure.

Those standards are intended to be consistent with the uranium fuel-cycle standards in 40 CFR Part 190 discussed above. Because the fuel-cycle standards were based primarily on judgments about doses that were reasonably achievable with available effluent-control technologies, rather than doses that must be achieved to protect public health, the difference between the standards for waste management and storage for facilities regulated by the Nuclear Regulatory Commission or Agreement States and the standards for DOE facilities not regulated by the Nuclear Regulatory Commission or Agreement

States is a clear example of establishing standards based on doses judged by EPA to be reasonably achievable.

The standards for disposal of spent fuel, high-level waste, and transuranic waste include the following provisions:

- For 10,000 years after disposal, cumulative releases of radionuclides to the accessible environment, taking into account inadvertent human intrusion (such as drilling) and undisturbed performance of the disposal system, shall have a likelihood of less than one chance in 10 of exceeding specified values and less than one chance in 1,000 of exceeding 10 times the specified values.

- For 10,000 years after disposal, undisturbed performance of the disposal system shall not cause the annual effective dose equivalent to individuals in the accessible environment from all potential exposure pathways to exceed 0.15 mSv (15 mrem).

- For 10,000 years after disposal, undisturbed performance of the disposal system shall not cause radioactivity in any underground source of drinking water in the accessible environment to exceed limits (that is, the MCLs) specified in 40 CFR Part 141 as they existed when the disposal standards became effective.

In the first of those provisions, referred to as the containment requirements, the limits on cumulative releases of radionuclides to the accessible environment were developed on the basis of judgments about releases that are reasonably achievable with foreseeable technology for disposal in geologic repositories. Thus, the containment requirements were based on an application of the ALARA objective. The release limits correspond to about 1,000 deaths in the US population over 10,000 y (an average of one every 10 y).

Naturally occurring radionuclides are important constituents of spent fuel and high-level waste but are unimportant in transuranic waste. For disposal of spent fuel and high-level waste, the containment requirements are expressed in terms of limits on cumulative releases of radionuclides per 1,000 metric tons of heavy metal (MTHM) irradiated in a reactor. For naturally occurring radionuclides, the specified release limits per 1,000 MTHM are 3.7 TBq (100 Ci) for ^{226}Ra, ^{234}U, ^{235}U, and ^{238}U; 0.37 TBq (10 Ci) for thorium-230 and thorium-232; and 37 TBq (1,000 Ci) for ^{210}Pb. Because the containment requirements apply only for 10,000 y, during which time the buildup of radium in chemically separated uranium would be relatively unimportant, and because thorium has been used only infrequently in nuclear fuel, the containment requirements should be more important for uranium than for other naturally occurring radionuclides.

In the second disposal requirement described above, the constraint on annual effective dose equivalent of 0.15 mSv (15 mrem) for individuals was based on three considerations. First, EPA judged that this dose corresponds to a limit on lifetime cancer risk that is consistent with the constraint on annual dose equivalents—0.25 mSv (25 mrem) to the whole body and 0.75 mSv (75 mrem) to any organ—in the original individual-protection requirement promulgated in 1985 (50 FR 38066). Second, EPA judged that this dose would provide an acceptable level of risk for the few individuals likely to be living near the small number of disposal sites. Third, EPA's analyses of the undisturbed performance of disposal systems (absent human intrusion) indicated that the specified dose constraint should be reasonably achievable at well-chosen sites.

The third disposal requirement described above addresses protection of groundwater near disposal sites. The essence of this requirement is that waste disposal should not cause any potential drinking-water supply to exceed standards (MCLs) developed under the Safe Drinking Water Act. This requirement reflects EPA's general policy that the nation's groundwater resources should be protected to avoid future costs of water treatment (EPA 1991b). In this regard, it is important to recall that drinking-water standards for radionuclides are based primarily on judgments about levels of radioactivity that are reasonably achievable given existing background and available technology for water treatment, rather than a judgment about risks to public health that must be achieved without regard for the costs of water treatment. Therefore, the groundwater-protection requirement for waste disposal clearly is not based solely on a judgment about acceptable risk posed by radionuclides in drinking water. EPA's analyses for undisturbed performance of waste-disposal systems also indicated that the groundwater-protection requirement should be reasonably achievable at well-chosen sites.

The containment requirements (cumulative release limits) in the disposal standards generated considerable controversy when they were promulgated in 1985, in part because they were based on estimated impacts on the entire US population but would not necessarily provide adequate protection of individuals near disposal sites. The individual-protection requirement promulgated in 1993 was intended to address that concern. However, the period of 10,000 y for applying the containment requirements also was controversial, given the much longer time over which spent fuel and high-level waste would remain hazardous. In the Energy Policy Act of 1992, Congress responded to both controversies by directing EPA to issue new standards, to be based on results of a study by the National Academy of Sciences, that would apply only to disposal of spent fuel and high-level waste at the Yucca Mountain site in Nevada, which is the only authorized disposal facility for such waste.

Furthermore, Congress indicated a preference for an individual-dose standard, rather than the existing containment requirements, for the Yucca Mountain site. The congressional directive essentially discarded the existing disposal standards for this site, especially the containment requirements.

The Academy report on Yucca Mountain standards was completed in 1995 (National Research Council 1995). The report recommended that the limits on cumulative releases of radionuclides to the accessible environment over 10,000 y be replaced with a standard for individual risk (not dose) that would be applied for a period consistent with the expected geologic stability of the Yucca Mountain site (perhaps on the order of 10^6 y).

EPA has not issued proposed new disposal standards for the Yucca Mountain site in response to the congressional directive. The only authorized facility to which the current disposal standards promulgated in 1993 apply is the Waste Isolation Pilot Plant (WIPP) in New Mexico for disposal of DOE's transuranic waste.

Standards for Low-Level Waste EPA has not issued proposed standards for management and disposal of low-level radioactive waste. In a draft proposed rule (EPA 1994a), EPA indicated a preference for a standard for management and storage in the form of a constraint on annual effective dose equivalent for individuals of 0.15 mSv (15 mrem) from all exposure pathways and a standard for disposal in the form of the same constraint on individual dose plus a separate requirement for groundwater protection that would be consistent with MCLs for radioactivity in drinking water established in 40 CFR Part 141 under the Safe Drinking Water Act. The preferred disposal standard thus would be consistent with the individual-protection and groundwater-protection requirements in the standards for disposal of spent fuel, high-level waste, and transuranic waste in 40 CFR Part 191. EPA also indicated a preference that the disposal standard for low-level waste apply for 1,000 y.

Standards for Cleanup of Radioactively Contaminated Sites

In addition to the standards for remediation of contaminated land and buildings at or near inactive uranium and thorium mill tailings sites, EPA develops standards for cleanup of other radioactively contaminated sites, including sites where deliberate disposal of radioactive waste occurred in the past. Any such standards apply to cleanup of naturally occurring radionuclides that were released to the environment or enhanced in the environment by human activities.

Current Cleanup Standards for Radionuclides Remediation of radioactively contaminated sites is regulated by EPA mainly under authority of CERCLA and its implementing regulations in the National Contingency Plan (NCP) in 40 CFR Part 300, which was promulgated in 1990 (55 FR 8666).

CERCLA addresses environmental contamination that is not properly regulated under other laws (such as the Atomic Energy Act, the Clean Air Act, the Safe Drinking Water Act, the Clean Water Act, and RCRA).

An essential feature of CERCLA and the NCP is that they do *not* specify a priori *requirements* for remediation of contaminated sites. That is, CERCLA and the NCP do not specify risks, doses, or levels of hazardous substances in the environment above which remedial actions are required regardless of cost or other circumstances. Rather, CERCLA and the NCP specify *preliminary* remediation *goals*, which normally are based on standards promulgated under other environmental laws. When the preliminary remediation goals are exceeded at a site, the feasibility of reducing risk must be investigated, but, as described later in this section, actions to reduce risk are not necessarily required. With the goals as a starting point, the cleanup levels actually achieved, as incorporated in the record of decision (ROD), are developed by a complex process of risk assessment, evaluation of the costs and benefits of alternatives for remediation, and negotiation among all stakeholders in the decision.

CERCLA and the NCP specify that the preliminary remediation goals at any site shall be protective of human health and the environment, and shall take into account:

- Applicable or relevant and appropriate requirements (ARARs) established under other federal or state environmental laws, with federal drinking-water standards established in 40 CFR Part 141 under authority of the Safe Drinking Water Act specified as ARARs for cleanup of contaminated groundwater and surface water.

- Other information to be considered (TBCs) that is not an ARAR, such as EPA's groundwater-protection strategy (EPA 1991b) and DOE Orders.

- For known or suspected carcinogens (such as radionuclides), an upper bound on lifetime cancer risk of 10^{-4} to 10^{-6} posed by all substances and all exposure pathways combined.

Several points about the preliminary remediation goals should be noted. First, for radionuclides, the drinking-water standards specified as ARARs for remediation of contaminated groundwater and surface water are the MCLs described earlier in this chapter, and the goals apply only to sources that are current or potential sources of drinking water. Second, TBCs generally are less important than ARARs and the cancer-risk criterion in developing remediation goals, because they have not been subjected to a public rule-making process. Finally, the goal for lifetime cancer risk was based on the levels of risk embodied in regulations established under other laws, including the Safe Drinking Water Act and the Clean Air Act. However, later EPA directives for carcinogens in general (Clay 1991) and radionuclides in particular (Luftig and Weinstock 1997) have indicated that lifetime cancer risks less than about 10^{-4} normally would not need to be considered in establishing preliminary remediation goals; that is, the cancer-risk goal of 10^{-4} to 10^{-6} specified in the NCP normally should be interpreted as a single value of 10^{-4}. Thus, cancer risks less than about 10^{-4} usually would be considered negligible because investigations into the feasibility of reducing risks beyond these low levels normally would not be required.

The preliminary remediation goals described above define desired levels of environmental contamination to be achieved in site cleanups under CERCLA. However, the goals are not unqualified requirements; CERCLA and the NCP also specify several conditions for waiving compliance with the ARARs, TBCs, or the cancer-risk goal in establishing actual cleanup levels at any site. Compliance with the preliminary remediation goals can be waived, for example, if the remedial action is an interim measure, if compliance would result in a greater risk to public health and the environment than noncompliance, if compliance is technically infeasible or impractical, if another response would achieve an equivalent level of protection, or if compliance would not balance the cost of the response against the benefit in protecting public health and the environment. In essence, compliance with the preliminary remediation goals is required only when it is practicable and cost-effective.

Negotiated cleanup levels at different sites, as incorporated in RODs, have varied considerably and usually have corresponded to lifetime cancer risks of about 10^{-4}-10^{-2}, that is, substantially above the goal of 10^{-4} (EPA 1994b; Baes and Marland 1989). Thus, lifetime risks above the goal of 10^{-4} clearly are not "unacceptable," because risks above the goal have been accepted by EPA and other stakeholders in most cleanup decisions. The most important factors in past cleanup decisions have been cost and feasibility, rather than compliance with ARARs, TBCs, or the cancer-risk goal. The process of arriving at negotiated cleanup levels at any CERCLA site thus clearly resembles applications of the ALARA objective to control of radiation exposures under authority of the Atomic Energy Act.

A few contaminated sites are being remediated under authority of RCRA rather than CERCLA. Although the definition of hazardous waste in RCRA specifically excludes radioactive materials as defined in the Atomic Energy Act (source, special nuclear, and byproduct materials), this exclusion does not apply to other radioactive materials, including naturally occurring radioactive materials, not associated with the nuclear fuel cycle. Of particular importance to site cleanups under RCRA are the provisions of EPA's implementing regulations in 40 CFR Part 264 that apply standards similar to the

MCLs for drinking water in 40 CFR Part 141 to protection of groundwater and require corrective actions if the groundwater-protection standards are exceeded.

Future Cleanup Standards for Radionuclides EPA intends to develop standards for cleanup of radioactively contaminated sites under authority of the Atomic Energy Act. These standards could be applied to licensees of the Nuclear Regulatory Commission or Agreement States, sites under control of a federal agency (such as DOE), and any sites subject to remediation under CERCLA that do not have signed RODs; but the standards probably will not be applied to facilities for disposal of spent fuel, high-level waste, or transuranic waste regulated under 40 CFR Part 191, uranium mill tailings sites that comply with 40 CFR Part 192, or sites that have been remediated under CERCLA with signed RODs.

In contrast with the current approach to remediation of contaminated sites under CERCLA—of first establishing preliminary remediation goals and then negotiating acceptable cleanup levels at specific sites that might be above the goals when compliance with the goals is not feasible—the cleanup standards for radionuclides that EPA intends to develop under the Atomic Energy Act would establish requirements that must be met to permit unrestricted or restricted release of contaminated sites. An important challenge for the new standards will be to reconcile the different concepts of risk goals under CERCLA and dose (risk) limits under the Atomic Energy Act and to show that the standards would be reasonably achievable at most sites.

EPA has not issued proposed standards for cleanup of radioactively contaminated sites, although it has performed technical analyses to address the costs and benefits of different cleanup levels (Wolbarst and others 1996). In a draft proposed rule (EPA 1994b) and a later directive (Luftig and Weinstock 1997), EPA has indicated a preference that a site could be released for unrestricted use if the annual effective dose equivalent to individuals, assuming a residential land-use scenario, would not exceed 0.15 mSv (15 mrem), which corresponds to a lifetime cancer risk of about 10^{-4}; if levels of radon in existing and future structures would comply with the guidance on indoor radon discussed later in this chapter; and if levels of radioactivity in groundwater that is a current or potential source of drinking water would comply with drinking-water standards (MCLs) in 40 CFR Part 141, unless compliance is technically impractical. Thus, the preferred standards for cleanup of radioactively contaminated sites would be consistent with existing standards for disposal of spent fuel, high-level waste, and transuranic waste in 40 CFR Part 191 and draft proposed standards for disposal of low-level waste discussed above. EPA also indicated that the cleanup standards for radionuclides should apply for 1,000 y.

Standards for Airborne Emissions of Radionuclides

Under authority of the Clean Air Act, EPA has established National Emission Standards for Hazardous Air Pollutants (NESHAPs) in 40 CFR Part 61. The current NESHAPs for radionuclides were first established in 1989 (54 FR 51654) and then amended in 1991 (56 FR 65934), 1992 (57 FR 23305), 1994 (59 FR 36280), 1995 (60 FR 46206), and 1996 (61 FR 68972). The current standards include the following provisions that apply to naturally occurring radionuclides:

- A limit on annual effective dose equivalent to individuals of 0.1 mSv (10 mrem) for DOE facilities emitting any radionuclide other than radon, except for disposal facilities subject to 40 CFR Part 191 or 40 CFR Part 192 and excluding the dose from ^{222}Rn and its decay products; non-DOE federal facilities, except for disposal facilities subject to 40 CFR Part 191, inactive uranium mill tailings disposal sites subject to 40 CFR Part 192, and low-energy accelerators, and excluding the dose from ^{222}Rn; and emissions of ^{222}Rn from specified underground uranium mines.

- A limit on emissions of polonium-210 from all calciners and nodulizing kilns at elemental-phosphorus plants of 0.07 TBq (2 Ci) per year or a limit on total emissions from any plant of 0.17 TBq (4.5 Ci) per year when specified scrubbers are installed.

- A limit on average concentration of ^{226}Ra of 0.4 Bq/g (10 pCi/g) in phosphogypsum distributed in commerce for uses in agriculture.

- A limit on emission rate of ^{222}Rn of 0.7 Bq/m^2 (20 pCi/m^2) per second from DOE facilities for storage and disposal of material containing radium, inactive phosphogypsum stacks (waste piles from phosphate mining), and operating and inactive uranium mill tailings piles, except for inactive disposal sites licensed by the Nuclear Regulatory Commission.

The standards issued in 1989 also applied to specified licensees of the Nuclear Regulatory Commission and Agreement States, including inactive uranium mill tailings disposal sites, nuclear power reactors, and facilities other than nuclear power reactors except for users of radionuclides only in the form of sealed sources. However, EPA has rescinded the standards for licensed commercial facilities on the basis of the 1990 Clean Air Act Amendments and a memorandum of understanding with the Nuclear Regulatory Commission. In effect, EPA has agreed that Nuclear Regulatory Commission regulations in 10 CFR Part 50, Appendix I, and 10 CFR Part 20 limit airborne emissions of radionuclides to an extent consistent with or more restrictive than EPA standards.

In establishing NESHAPs for radionuclides, EPA followed a previous mandate by the court of appeals regarding the standard for vinyl chloride (for example, EPA 1989d). The court directed EPA to use a two-step decision process in setting NESHAPs. Specifically, EPA was to determine a "safe" or "acceptable" risk to individuals or populations and an "ample margin of safety" below the safe or acceptable risk for protection of public health. The court ruled that technical feasibility and cost could not be the primary basis for the standards, as has often been the case with EPA standards developed under other laws, including the Atomic Energy Act, the Safe Drinking Water Act, and the Clean Water Act.

In response to the court order, EPA set NESHAPs for radionuclides (and other carcinogens) such that the lifetime risk to individuals would not exceed about 10^{-4} and the lifetime risk to the greatest number of individuals in exposed populations (that is, the average individual risk) would not exceed about 10^{-6} (EPA 1989d). The limits on risk used in establishing the NESHAPs were developed on the basis of a survey of other societal risks, and the assumed acceptable risks were consistent with other regulatory precedents, such as standards (MCLs) for radionuclides and other carcinogens in drinking water in 40 CFR Part 141.

Several additional points about the approach to setting NESHAPs, which applied before the 1990 Clean Air Act Amendments, should be noted. First, radionuclides and other carcinogens were regulated individually. Although the NESHAPs for any carcinogens were based primarily on considerations of acceptable risk, there was no standard defining a limit on acceptable risk posed by exposure to airborne emissions of all carcinogens combined.

Second, EPA did not establish NESHAPs for some sources of airborne emissions of naturally occurring radionuclides, including surface uranium mines and coal-fired boilers. In both cases, the estimated risks to maximally exposed and average individuals were less than the assumed acceptable levels of 10^{-4} and 10^{-6}, respectively, although the estimated average risk posed by coal-fired boilers was higher than the average risk associated with DOE facilities or nuclear power plants (EPA 1989d). In addition, releases of radionuclides for those two source categories are effectively controlled under other laws and regulations (EPA 1989d).

Third, in establishing the standards for radionuclides, especially the authorized limit on annual effective dose equivalent of 0.1 mSv (10 mrem) for many sources, EPA ignored a statement by NCRP (1984a) that a limit on annual effective dose equivalent of 1 mSv (100 mrem) from all controlled sources

GUIDANCES AND REGULATIONS FOR NORM 135

combined provides an upper bound on acceptable risk to individuals and that an authorized limit for specific sources or practices at 25% of the dose limit for all sources and further application of the ALARA objective at specific sites would provide an adequate margin of safety for exposed individuals and populations.

Fourth, EPA envisioned that the two-step decision process based on considerations of acceptable risk, as mandated by the court of appeals in the vinyl chloride case, would be applied only in establishing NESHAPs under the Clean Air Act but would not be applied to any other regulations developed under the act or to regulations developed by EPA under any other laws (such as the Atomic Energy Act and the Safe Drinking Water Act).

Finally, although the court of appeals mandated considerations of acceptable risk, the resulting standards were shown to be reasonably achievable with existing effluent-control technologies (EPA 1989d). EPA (1989d; 1989b) also noted that the feasibility of emission controls was considered in determining an ample margin of safety, which is the second part of the two-step decision process described above and was used in establishing the limit on average individual risk of about 10^{-6}.

In the Clean Air Act Amendments of 1990, the risk-based approach to setting NESHAPs was replaced with an approach based primarily on maximum achievable control technology. However, all NESHAPs established before the Amendments Act, including the standards for radionuclides described in this section, remain in effect.

Guidance on Radon in Homes

In 1986, EPA and the Department of Health and Human Services issued guidance on radon in homes (EPA and DHHS 1986), which poses the greatest radiation risk to the public (see table 2.10). The guidance included the statement that mitigation of exposures was indicated for radon concentrations above 150 Bq/m^3 (4 pCi/L), which corresponds to an exposure to short-lived radon decay products of about 4×10^{-7} J/m^3 (0.02 WL). The recommended mitigation level was based on considerations of risks to individuals, an analysis of existing levels of radon in homes, and the costs and benefits of reducing these levels. The guidance was not a standard for limiting exposures of the public to radon in homes, but it has been widely used in the real-estate and home-insurance industries.

The Indoor Radon Abatement Act of 1988 established the goal of reducing indoor radon concentrations to background (outdoor) levels, which average about 7 Bq/m^3 (0.2 pCi/L) but are highly variable (NCRP 1987a). In response to the act, the guidance on indoor radon was reevaluated (EPA and DHHS 1994). The guidance continues to state, on the basis of further cost-benefit analysis, that radon concentrations in homes above 150 Bq/m^3 (4 pCi/L) indicate a need for mitigation of exposures. However, the current guidance also recommends that mitigation of exposures be considered for concentrations of 70-150 Bq/m^3 (2-4 pCi/L), especially if the concentrations can be reduced below 70 Bq/m^3 (2 pCi/L).

For exposure at the recommended mitigation level of 150 Bq/m^3 (4 pCi/L), the estimated lifetime risk of fatal lung cancers is 2×10^{-3} for people who have never smoked and 3×10^{-2} for smokers (EPA and DHHS 1994). For former smokers, the risk might lie between those two values.

EPA guidance on indoor radon is considered in more detail in chapter 8.

Applicability of EPA Guidances and Regulations to TENORM

Up to now, this chapter has discussed EPA's published guidances and regulations, either existing or proposed, that apply to naturally occurring radionuclides without regard for whether the standards apply to TENORM, as defined in this study, or to naturally occurring radionuclides associated with operations of the nuclear fuel cycle, which are not included in TENORM. The following statements summarize the applicability of the various guidances and regulations to TENORM, except the guidance for indoor radon, which clearly is concerned only with a particular type of TENORM.

- Existing federal guidance on radiation protection of the public and EPA's proposed revision of the federal guidance are intended to apply to all sources of exposure to TENORM, except indoor radon.

- Standards for operations of uranium fuel-cycle facilities in 40 CFR Part 190 do not apply to TENORM, because they apply only to radioactive materials regulated under the Atomic Energy Act.

- Standards for radioactivity in drinking water in 40 CFR Part 141 apply to TENORM from any source (and also include natural background).

- Standards for liquid discharges from mines or mills used to produce or process uranium, radium, and vanadium ores in 40 CFR Part 440 apply to TENORM from the specified sources.

- Standards for uranium and thorium mill tailings in 40 CFR Part 192 do not apply to TENORM, because they apply only to radioactive materials regulated under the Atomic Energy Act. However, because mill tailings contain only naturally occurring radionuclides, the standards have been widely used as a model for regulating TENORM (see chapter 9).

- Standards for management and disposal of spent fuel, high-level waste, and transuranic waste in 40 CFR Part 191 and standards for management and disposal of low-level waste that might be developed do not apply to TENORM, because they apply only to radioactive materials regulated under the Atomic Energy Act.

- Standards for cleanup of radioactively contaminated CERCLA sites in 40 CFR Part 300 apply to TENORM.

- Standards for airborne emissions of radionuclides in 40 CFR Part 61 apply to TENORM from the specified sources.

In chapter 10, EPA guidances and regulations that apply to TENORM are summarized and compared with guidances for TENORM developed by other organizations.

Other EPA Initiatives for TENORM

In addition to the guidances and regulations for TENORM discussed previously in this chapter, EPA has undertaken other initiatives for TENORM. These include the development of guidelines for disposal of wastes arising from treatment of drinking water and for the use and disposal of sewage sludge.

Guidelines for Drinking-Water Treatment Wastes EPA has developed suggested guidelines for disposal of drinking-water treatment wastes that contain naturally occurring radionuclides (EPA 1994e). The current guidelines supersede those issued previously (EPA 1990). The guidelines are intended only to provide assistance to drinking-water treatment facilities where gaps in existing state regulations for disposal of wastes containing naturally occurring radionuclides exist, but they do not establish or affect any legal rights or obligations. Separate guidelines were developed for disposal of liquid and solid wastes.

The guidelines for disposal of liquid wastes from treatment of drinking water consider disposal into storm sewers, surface waters, sanitary sewers, and wells. These guidelines are summarized as follows.

For disposal into storm sewers and surface waters, requirements for obtaining National Pollutant Discharge Elimination System (NPDES) permits established under the Clean Water Act generally apply. That is, releases of liquid wastes from treatment of drinking water that contains naturally occurring radionuclides are subject to limits specified in NPDES permits, and no additional guidance is needed.

For disposal into sanitary sewers, the suggested guidelines include the following:

- The daily quantities of soluble ^{226}Ra and ^{228}Ra, diluted by the average daily quantity of water-treatment wastes released into the sewer, should not exceed 15 Bq/L (400 pCi/L) and 30 Bq/L (800 pCi/L), respectively.

- The daily quantity of soluble natural uranium, diluted by the average daily quantity of water-treatment wastes released into the sewer, should not exceed 37 kBq/L (1 µCi/L).

- The gross quantity of radioactive material released by a facility into the sanitary sewer should not exceed 37 GBq (1 Ci) per year.

Those guidelines were based on standards for Nuclear Regulatory Commission and Agreement State licensees that had been established in 10 CFR Part 20.

The guidelines for subsurface disposal in wells were based on regulations for the Underground Injection Control (UIC) program established under the Safe Drinking Water Act. The UIC program distinguishes between radioactive and nonradioactive wastes on the basis of radionuclide concentrations, and the concentration limits for nonradioactive waste for the naturally occurring radionuclides of concern are 1.1 Bq/L (30 pCi/L) for ^{226}Ra and ^{228}Ra and 1.1 kBq/L (30,000 pCi/L) for natural uranium. The suggested guidelines include the following:

- Shallow injection of radioactive waste—that is, injection above or into an underground source of drinking water (USDW)—is banned under the UIC program.

- Deep-well disposal of radioactive waste below a USDW or shallow injection of nonradioactive waste is considered a Class V well, but no recommendations are made regarding disposal of drinking-water treatment wastes that contain naturally occurring radionuclides in Class V wells.

- Nonradioactive waste should be disposed of in a Class I well beneath the lowest USDW.

If there are no acceptable methods for disposal of liquid wastes, then the wastes normally should be solidified for disposal in accordance with the guidelines for solid waste arising from water treatment.

The guidelines for disposal of solid drinking-water treatment wastes that contain naturally occurring radionuclides also depend on the concentrations of radionuclides. The guidelines are as follows.

- Solid wastes that contain ^{226}Ra plus ^{228}Ra at less than 0.11 Bq/g (3 pCi/g) and uranium at less than 50 µg/g may be disposed of, without the need for long-term institutional controls, in a municipal landfill, provided that the radioactive wastes are mixed with other materials when emplaced and that the radioactive wastes constitute less than about 10% of the volume of material in the landfill.

- Solid wastes that contain ^{226}Ra plus ^{228}Ra at 0.11-1.9 Bq/g (3-50 pCi/g) should be disposed of with a cover that would protect against radon release and would isolate the wastes, and institutional controls designed to avoid inappropriate uses of the disposal site should be provided. Sites that comply with disposal requirements for nonhazardous waste developed under Subtitle D of RCRA would be adequate.

- For solid wastes that contain ^{226}Ra plus ^{228}Ra at 1.9-74 Bq/g (50-2,000 pCi/g), the disposal method should be determined case-by-case. The disposal options considered should include methods that comply with standards for disposal of uranium mill tailings or with standards for disposal of hazardous waste developed under Subtitle C of RCRA, a facility licensed by a state for waste that contains naturally occurring and accelerator-produced radioactive material (NARM), and for concentrations approaching 74 Bq/g (2,000 pCi/g) a facility for low-level radioactive waste licensed by the Nuclear Regulatory Commission or an Agreement State under the Atomic Energy Act or a facility permitted by EPA or a state to dispose of discrete NARM.

- For solid wastes that contain uranium at 50-500 µg/g, the disposal method should be determined case-by-case. Disposal at sites licensed by states for NARM waste or other radioactive wastes and recovery of the uranium when the wastes contain uranium at more than 0.05% by weight should be considered.

- Wastes that contain radium at more than 74 Bq/g (2,000 pCi/g) or uranium at more than 500 µg/g should be disposed of only as permitted by regulations.

- For wastes containing ^{210}Pb, the disposal practice should be based on case-by-case reviews.

The concentration limits of 0.11 Bq/g (3 pCi/g) for ^{226}Ra plus ^{228}Ra and 50 µg/g for uranium for disposal in municipal landfills without the need for physical barriers or long-term institutional controls were based on the principle that the relatively high average risks posed by exposure to background radium and uranium in soil should not be allowed to increase by more than a small amount.

EPA's suggested guidelines for disposal of radioactive waste arising from treatment of drinking water are not considered further in this report.

Radioactivity in Sewage Sludge In 1993, EPA established standards in 40 CFR Part 503 for the use or disposal of sewage sludge (58 FR 9248), including standards for selected heavy metals and pathogens. However, standards for radioactivity are not included.

During this study, the committee was informed of current work by the Interagency Steering Committee on Radiation Standards to investigate levels of radioactivity in sewage sludge and the need for appropriate guidance. Its Subcommittee on Sewage Sludge and Ash is conducting a survey of municipal sewage treatment plants to determine levels of TENORM and NRC licensee discharged radionuclides found in sludge and ash.

Other Alternatives for EPA Regulation of TENORM

As noted earlier, EPA also may regulate NARM and therefore TENORM, which are not subject to regulation under the Atomic Energy Act, under TSCA and RCRA. Although EPA has not developed any such regulations, this section briefly considers the possible regulation of TENORM under TSCA and RCRA.

Regulation of TENORM under the Toxic Substances Control Act TSCA is concerned with protection of human health and the environment in the use of toxic substances in commerce. Source, special nuclear, and byproduct materials regulated under the Atomic Energy Act are excluded from regulation under TSCA, but the exclusion does not apply to NARM. Therefore, EPA may establish standards for management and disposal of TENORM under TSCA if the unregulated use and disposal of these materials presents an unreasonable risk of injury to health or the environment (Cameron 1996; EPA 1989a).

In 1989, EPA prepared an unpublished draft standard for NARM under TSCA (EPA 1989a). It applied only to the relatively small volumes of material that contains radioactivity at more than 74 Bq/g (2 nCi/g), that is, to so-called discrete sources. EPA indicated its intention that discrete NARM and TENORM should be regulated as though they were low-level radioactive waste, which is regulated under the Atomic Energy Act. The draft standard did not indicate how EPA intended to regulate the much larger volumes of TENORM that contain lower concentrations of radionuclides, that is, the so-called diffuse sources.

Regulation of TENORM under the Resource Conservation and Recovery Act RCRA is concerned, in part, with management and disposal of hazardous and nonhazardous solid waste. As in the case of TSCA, radioactive materials regulated under the Atomic Energy Act are excluded from regulation under RCRA, but the exclusion does not apply to NARM. Thus, EPA may establish standards for management and disposal of waste that contains TENORM under RCRA. An important distinction between regulation of TENORM under TSCA and under RCRA is that RCRA applies only to waste materials, whereas TSCA applies to uses of materials, as well as wastes. As described below, TENORM could be regulated under RCRA in two ways.

First, waste that contains TENORM could be regulated under Subtitle C of RCRA, which addresses management and disposal of solid hazardous waste. However, the definition of hazardous waste in EPA regulations that implement Subtitle C of RCRA does not include NARM or, therefore, TENORM, and EPA cannot regulate TENORM as hazardous waste unless it is so declared in regulations that implement RCRA (40 CFR Part 261). Furthermore, some potentially important wastes that contain TENORM are specifically excluded from the current definition of hazardous waste in 40 CFR Part 261, including mining overburden that is returned to the mine site; some wastes generated from the combustion of coal or other fossil fuels; wastes associated with the exploration, development, or production of crude oil, natural gas, or geothermal energy; and solid waste from the extraction, beneficiation, and processing of some ores and minerals, including coal, phosphate rock, and overburden from the mining of uranium ore. Therefore, for EPA to regulate waste that contains TENORM under Subtitle C of RCRA, substantial changes in the current regulatory definition of solid hazardous waste would be required.

As an alternative, waste that contains TENORM could be regulated under Subtitle D of RCRA, which is concerned with disposal of nonhazardous waste in municipal (sanitary) landfills. This approach has the advantage that changes in the regulatory definition of hazardous waste to include TENORM would not be required. However, it would be appropriate only if disposal of waste that contains TENORM in the same manner as ordinary household trash would provide adequate protection of public health and the environment. Therefore, disposal of waste that contains TENORM as nonhazardous waste under RCRA presumably would be suitable only for materials that contain relatively low concentrations of radionuclides that would pose no more than negligible risks to public health and the environment.

Risks Corresponding to EPA Guidances and Regulations

An additional perspective on EPA's guidances and regulations for naturally occurring radionuclides discussed above and summarized in table 7.1 is provided by the data in table 7.2, where quantitative criteria in various guidances and regulations are expressed in terms of the corresponding lifetime risk of fatal cancers, assuming continuous exposure over 70 y. Table 7.2 also includes the lifetime risks resulting from exposure to natural background radiation, including indoor radon, and to indoor radon only. In converting effective dose equivalents to risk, the risk per unit dose is assumed to be 5×10^{-5} per millisievert (5×10^{-7} per millirem) (EPA 1994c; NCRP 1993a; ICRP 1991), and the risk corresponding to a specified dose to a particular organ takes into account the relationship between organ dose equivalent and effective dose equivalent for the particular radionuclide of concern, as obtained from current federal guidance (Eckerman and others 1988).

The data in table 7.2 indicate that the risks corresponding to the various guidances and regulations that apply to naturally occurring radionuclides vary considerably and that many of the risks corresponding to current standards are considerably smaller than the risks posed by natural background radiation. However, as discussed in the following section, there are many valid reasons why the various guidances and regulations are not consistent with regard to the corresponding levels of risk.

CONSISTENCY OF DIFFERENT GUIDANCES AND REGULATIONS

As the number of environmental laws and regulations has increased in recent years, there has been considerable interest in the issue of the consistency of standards, that is, the extent to which quantitative criteria contained in the different guidances and regulations for radionuclides and hazardous chemicals in the environment correspond to similar health risks to the public (for example, Overy and Richardson 1995; Taylor 1995; GAO 1994; Brown 1992; EPA-SAB 1992; Kocher and Hoffman 1992; Remick 1992; Kocher and Hoffman 1991; Kocher 1988; Travis and others 1987). In discussing the risks corresponding to

Table 7.2. Lifetime cancer risks corresponding to selected radiation exposures and EPA guidances and regulations for controlling exposures of the public[a]

Risk	Exposure or guidance or regulation
4×10^{-2}	Mill tailings standards (cleanup of contaminated land and buildings)
$0.2\text{-}3 \times 10^{-2}$	Concentration of radon in homes of 150 Bq/m^3 (EPA and DHHS 1994)[b]
2×10^{-2}	Annual dose equivalent to whole body from external exposure to all controlled sources combined of 5 mSv (existing FRC guidance)
1×10^{-2}	Average annual effective dose equivalent from exposure to natural background radiation, including indoor radon, of 3 mSv (NCRP 1987a)
$0.7\text{-}9 \times 10^{-3}$	Average indoor radon concentration of 50 Bq/m^3 (EPA and DHHS 1994)[b]
4×10^{-3}	Annual effective dose equivalent from all controlled sources combined, excluding indoor radon, of 1 mSv (proposed federal guidance)
4×10^{-3}	Indoor gamma radiation level of 20 µR/h and indoor residence time of 85%
2×10^{-3}	Concentrations of ^{226}Ra in soil of 0.2 Bq/g in top 15 cm and 0.6 Bq/g below 15 cm and continuous external exposure indoors and outdoors
9×10^{-4}	Annual dose equivalent to whole body of 0.25 mSv
5×10^{-4}	Annual effective dose equivalent of 0.15 mSv
4×10^{-4}	Annual effective dose equivalent of 0.1 mSv
2×10^{-4}	Concentration of uranium in drinking water of 20 µg/L
2×10^{-4}	Concentration of ^{226}Ra in drinking water of 0.2 Bq/L
1×10^{-4}	Goal for cleanup of radioactively contaminated sites (CERCLA and NCP)
1×10^{-4}	Annual effective dose equivalent of 0.04 mSv (proposed drinking-water standard for beta- or gamma-emitting radionuclides)

Table 7.2. (continued)

Risk	Exposure or guidance or regulation
1×10^{-4}	Annual dose equivalent to lungs from inhalation of insoluble natural uranium of 0.25 mSv (uranium fuel-cycle standards)
4×10^{-5}	Annual dose equivalent to bone surfaces from ingestion of soluble natural uranium of 0.25 mSv (uranium fuel-cycle standards)
3×10^{-8}	Containment requirements for disposal of spent fuel, high-level waste, and transuranic waste (average risk in US population)

[a]Values assume continuous exposure over 70 y and, unless otherwise noted, risk of fatal cancers per unit effective dose equivalent of 5×10^{-5} per millisievert (EPA 1994c; NCRP 1993a; ICRP 1991).
[b]Lower bound for risk applies to individuals who have never smoked, and upper bound applies to smokers; for former smokers, risk may lie in between.

different guidances and regulations, various investigators have developed tables similar to table 7.2 (for example, GAO 1994; Kocher 1988; Travis and others 1987). As noted earlier, the comparisons have indicated that the risks corresponding to different guidances and regulations vary considerably, and some investigators have concluded that a consensus on acceptable risk is lacking (GAO 1994).

The desire for consistency in regulations is understandable. However, several factors indicate that it is unreasonable to expect it, including differences in statutory and judicial mandates, differences in the primary bases of standards, differences in the applicability of standards, differences in population groups of primary concern, and differences in considerations of natural background. The discussions of these and other factors in the following sections are concerned with guidances and regulations for radiation exposure of the public, but some also apply to regulation of hazardous chemicals.

Differences in Statutory and Judicial Mandates

A fundamental reason why the health risks corresponding to some of the guidances and regulations for radionuclides in the environment appear to be inconsistent is that the standards were developed under different laws that mandate different approaches to standard-setting. In particular, the traditional approach to establishing radiation standards under authority of the Atomic Energy Act is fundamentally different from the approach used in establishing radiation standards under the authority of other laws that are concerned primarily with exposures to hazardous chemicals (Overy and Richardson 1995; Kocher and Hoffman 1992, 1991).

The Atomic Energy Act provides the authority for regulation of radiation exposures of the public that arise, either directly or indirectly, from operations of the nuclear fuel cycle for peaceful or defense purposes. The traditional approach to radiation protection under the Atomic Energy Act has the following two basic elements, given that the exposures are justified:

- A *limit* (upper bound) on acceptable dose (and therefore risk), meaning that doses above the limit are regarded as *intolerable*.

- *Reduction* of doses (and risks) below the limit to as low as reasonably achievable (ALARA).

Those elements are an essential aspect of current recommendations on radiation protection developed by the ICRP (1991) and NCRP (1993a) and they are embodied in EPA's proposed federal guidance on radiation protection of the public (EPA 1994d), which applies to all controlled sources combined except indoor radon and medical exposures, and in regulations for specific sources or practices, including operations of uranium fuel-cycle facilities (40 CFR Part 190) and management and disposal of spent fuel, high-level waste, and transuranic waste (40 CFR Part 191).

The approach to controlling radiation exposures of the public under the authority of other environmental laws is, in many cases, quite the opposite of the approach to radiation protection under the Atomic Energy Act described above (Kocher and Hoffman 1992, 1991). Specifically, the approach under other laws often has the following two basic elements:

- A *goal* for acceptable risk.

- Allowance for an *increase* (relaxation) in risks above the goal on the basis, for example, of technical feasibility and cost.

Those elements are embodied, for example, in the requirements of the Safe Drinking Water Act and CERCLA and their implementing regulations. The Safe Drinking Water Act essentially sets a goal of zero risk to the public posed by exposure to radionuclides and other carcinogens in drinking water, but the goal clearly cannot be achieved at any cost. The act then requires that the legally enforceable standards (MCLs) be set as close to the goal of zero risk as possible, with technical feasibility and costs of removing radionuclides from public drinking-water supplies taken into account. The requirements of CERCLA and its implementing regulations (40 CFR Part 300) include compliance with ARARs and a lifetime cancer risk of 10^{-4} as goals for remediation of contaminated sites (Luftig and Weinstock 1997; Clay 1991), but these goals can be relaxed on the basis of many considerations, including that achieving the goals is not feasible.

It cannot be overemphasized that the concept of a *limit*, as embodied in radiation protection standards for the public developed under the Atomic Energy Act, is fundamentally different from the concept of a *goal*, as embodied in radiation standards developed under some other environmental laws. A goal for acceptable risk does not define any kind of a limit on acceptable (tolerable) risk that must be met without regard for cost or other relevant factors. Therefore, it is potentially misleading, and could be inappropriate, to compare quantitative criteria in the form of limits with criteria that are goals. For example, it is not particularly meaningful to compare the limit on lifetime cancer risk of about 4×10^{-3} corresponding to the primary dose limit of 1 mSv (100 mrem) per year for exposure over 70 years in EPA's proposed federal guidance on radiation protection of the public with the risk goal of 10^{-4} for cleanup of contaminated sites under CERCLA unless the fundamental difference in concept between the two is recognized.

An example of the importance of a judicial mandate in establishing standards is provided by the standards for airborne emissions of radionuclides developed under the Clean Air Act. The court of appeals mandated that the standards be based on considerations of acceptable risks to the public, whereas other standards for specific sources or practices developed by EPA have been based primarily on considerations of the achievability of risks (cost-benefit analysis). However, the standards developed under the Clean Air Act are reasonably consistent with most other standards that were based on the achievability of risks, in part because the lifetime risks of about 10^{-4} to 10^{-6} judged by EPA to be acceptable for airborne emissions also were reasonably achievable and because the risks judged by EPA to be acceptable for airborne emissions were comparable with the risks corresponding to other standards that were based on the achievability of risks.

Differences in Primary Bases of Standards

As discussed earlier, some radiation standards were based primarily on judgments about acceptable health risks to the public, and others primarily on judgments about the achievability of risks. There is no a priori reason to expect that risks corresponding to the two types of standards would be consistent.

The importance of the different bases of standards is illustrated by a comparison of EPA's proposed federal guidance on radiation protection of the public with the standards for radionuclides in drinking water. As indicated in table 7.2, the drinking-water standards for naturally occurring radionuclides correspond to lifetime risks of about 10^{-4}, whereas the primary dose limit of 1 mSv (100 mrem) per year in the proposed federal guidance corresponds to a lifetime risk of about 4×10^{-3}. The primary dose limit is based on an assumption about the maximum acceptable (tolerable) risk posed by radiation exposure whereas the drinking-water standards (MCLs) are based essentially on a cost-benefit analysis of removal of radionuclides from public drinking-water supplies. In general, standards based primarily on risks judged to be acceptable should not be compared with standards based primarily on risks judged to be reasonably achievable unless the difference between the two concepts is recognized.

Differences in Applicability of Standards

In many cases, the risks corresponding to various guidances and regulations appear to be inconsistent essentially because the standards differ in their applicability. Some of the ways in which differences in the applicability of standards are important are discussed below.

Perhaps the most important difference in the applicability of standards is shown by a comparison of EPA's proposed federal guidance on radiation protection of the public—specifically, the primary dose limit of 1 mSv (100 mrem) per year, which applies to all controlled sources of exposure combined except for indoor radon and medical exposures—with any other EPA guidances or regulations developed under any environmental laws, which apply only to specific sources or practices. A standard for all sources of exposure combined is not directly comparable with standard for a specific practice or source. Indeed, except for indoor radon, the risks corresponding to standards for specific sources or practices should be substantially less in most cases than the risk corresponding to the primary dose limit for all sources combined (EPA 1994d). In this regard, it should be noted that no guidance or regulation for hazardous chemicals specifies a limit on risk posed by exposure to all controlled sources combined. That is, for hazardous chemicals, there is no standard analogous to the primary dose limit in radiation-protection standards; rather, all standards for hazardous chemicals in the environment apply only to specific exposure situations. Furthermore, for any particular situation (such as contaminants in drinking water or airborne emissions of hazardous air pollutants), hazardous chemicals often have been regulated only individually.

A second important difference is that the various standards for specific sources or practices apply to different exposure situations. Most standards for specific sources or practices were based primarily on judgments about environmental levels, releases, or doses (and therefore risks) that are reasonably achievable for the exposure situations of concern (application of the ALARA objective). There is no a priori reason to expect risks judged reasonably achievable for one exposure situation (such as releases from operating nuclear facilities) to be consistent with risks judged reasonably achievable for a different situation (such as radioactive waste disposal). Indeed, it is primarily in the interest of achieving some degree of consistency in regulation that the quantitative criteria contained in standards that apply to different exposure situations often are about the same.

A third important difference is that standards developed under the Atomic Energy Act generally apply to all release and exposure pathways combined for the exposure situations of concern, whereas standards developed under other environmental laws often apply only to particular release and exposure pathways. For example, the dose constraint for operations of uranium fuel-cycle facilities (40 CFR Part 190) developed under the Atomic Energy Act applies to all release and exposure pathways, whereas standards for radioactivity in drinking water developed under the Safe Drinking Water Act (40 CFR Part 141) apply only to a single environmental medium (water) and a single exposure pathway, and standards developed under the Clean Air Act (40 CFR Part 61) apply only to airborne releases. The one exception for standards developed under laws other than the Atomic Energy Act is the cancer-risk goal of 10^{-4} for remediation of contaminated sites under CERCLA (40 CFR Part 300); in this case, the goal applies to all release and exposure pathways combined at a particular site. In general, it should not be expected that the risks corresponding to standards that apply only to a single release or exposure pathway would be consistent with the risks corresponding to standards that apply to all release and exposure pathways combined.

Differences in Population Groups of Primary Concern

Some standards are concerned primarily with protection of maximally exposed individuals; others are concerned primarily with protection of whole populations, that is, individuals in the population receiving an average exposure. For a given exposure situation, doses and risks for maximally exposed individuals generally will be higher than those for average individuals in the population. Therefore, the standards might differ substantially depending on the population group of primary concern.

Examples of standards that are concerned primarily with protection of maximally exposed individuals include the dose constraints in standards for operations of uranium fuel-cycle facilities (40 CFR Part 190) and management and disposal of spent fuel, high-level waste, and transuranic waste (40 CFR Part 191). Another example is the risk goal of 10^{-4} in standards for cleanup of contaminated sites under CERCLA (40 CFR Part 300). The standards for airborne emissions of radionuclides (40 CFR Part 61) took into account both the maximum individual risk and the average risk in the exposed population, but the dose constraint that applies to many sources is concerned primarily with protection of maximally exposed individuals.

The clearest example of a standard that is concerned with protection of whole populations, rather than maximally exposed individuals, is the containment requirements for disposal of spent fuel, high-level waste, and transuranic waste (40 CFR Part 191). The limits on cumulative releases of radionuclides over 10,000 y were based on an assumed number of health effects in the entire US population, without regard for risks to individuals who might reside near disposal sites, which are limited by a separate dose constraint. The drinking-water standards for radionuclides (40 CFR Part 141) also are concerned with protection of whole populations because the standards were derived on the basis of a cost-benefit analysis in which all individuals were assumed to ingest the same amount of drinking water.

Another example of the importance of the population group of concern in establishing standards is provided by current guidances on mitigation of radon in homes, specifically the EPA action level of 150 Bq/m^3 (4 pCi/L) compared with the NCRP-recommended action level of about 370 Bq/m^3 (10 pCi/L) discussed in chapter 8. EPA and NCRP both were concerned with mitigation of risks to the relatively few individuals who reside in homes in which the levels of radon greatly exceed the US average. However, the two organizations arrived at different action levels largely because EPA also was concerned with reduction of exposures in the greatest number of homes, and EPA developed its action level on the basis of a cost-benefit analysis for reduction of levels of indoor radon in all homes.

Differences in Considerations of Natural Background

In some cases, the health risks corresponding to various guidances and regulations appear to be inconsistent essentially because some standards are concerned with exposures to naturally occurring radionuclides and others are not. Given the relatively high doses and risks posed by exposure to natural background radiation (see chapter 2), the risks corresponding to various guidances and regulations can differ substantially depending on whether the standards include exposures to natural background.

The clearest examples of the importance of natural background in establishing standards are the regulations for control and cleanup of residual radioactive materials at uranium and thorium mill tailings sites (40 CFR Part 192) and the federal guidance on indoor radon (EPA and DHHS 1994). Both standards are concerned with exposures to naturally occurring radionuclides that have been increased by human activities, and knowledge of background levels of naturally occurring radionuclides was important in developing the standards. In either case, background levels result in relatively high doses and risks, and it clearly is impractical for the standards to require reductions in concentrations to levels below background. Therefore, it is reasonable that the risks corresponding to the mill tailings standards and the guidance on indoor radon are considerably higher than the risks corresponding to other standards that do include contributions from natural background, such as standards for operations of uranium fuel-cycle facilities (40 CFR Part 190) and waste management and disposal (40 CFR Part 191).

Other Considerations in Comparing Standards

Two additional factors have resulted in differences in risks corresponding to various guidances and regulations for controlling radiation exposures of the public.

First, the various guidances and regulations were developed at different times, and judgments about the acceptability of doses and risks have changed considerably over time. For example, when the standards for operations of uranium fuel-cycle facilities (40 CFR Part 190) were developed in the middle 1970s, the primary dose limit for all controlled sources combined was 5 mSv (500 mrem) per year (FRC 1961; FRC 1960), the risk of fatal cancers was assumed to be about 1×10^{-5} per millisievert (ICRP 1977), and standards for radionuclides and hazardous chemicals developed under environmental laws other than the Atomic Energy Act had not been issued or did not yet have an influence on radiation standards developed under the Atomic Energy Act. Since then, the primary dose limit for all controlled sources combined has been reduced to 1 mSv (100 mrem) per year, the assumed risk of fatal cancers has increased to 5×10^{-5} per millisievert (EPA 1994c; NCRP 1993a; ICRP 1991), and a judgment by EPA that a lifetime risk of about 10^{-4} is an upper bound on acceptable risk for specific sources or practices has been increasingly incorporated into radiation standards on the basis of precedents in regulations developed under other environmental laws (such as the Safe Drinking Water Act, the Clean Air Act, CERCLA). Thus, there has been a tendency in recent years to develop increasingly stringent radiation standards, as illustrated by EPA's use of a dose constraint of 0.15 mSv (15 mrem) or 0.1 mSv (10 mrem) per year, in contrast with the earlier use of a dose constraint of 0.25 mSv (25 mrem) per year.

Second, the dosimetric quantities used in radiation standards have changed over time. The earliest standards were expressed in terms of dose to the whole body or the critical organ. A weakness of this approach is that the dose criteria generally do not correspond well to a particular risk, especially for nonuniform irradiations of the body. However, later standards are expressed in terms of the effective dose equivalent, which was intended to be proportional to risk for any uniform or nonuniform irradiations of the body (ICRP 1977). The differences between organ doses and the effective dose equivalent are important mainly for ingestion and inhalation exposures. For most radionuclides, the effective dose equivalent per unit activity intake is substantially less than the dose to the critical organ; furthermore, the ratio of the two doses depends on the particular radionuclide (Eckerman and others 1988). But those differences are important only if dose criteria are compared; they are not important when the corresponding risks are compared, provided that conversion of organ dose to risk takes into account the dose in all tissues irradiated.

Summary of Issues of Consistency of Standards

There are several important reasons why the risks corresponding to the many guidances and regulations for controlling radiation exposures of the public appear to be inconsistent and why it is unreasonable to expect the risks to be consistent. The considerable variability in risks embodied in the various guidances and regulations is explained in large part by differences in legislative and judicial mandates for setting standards, differences in the primary bases of standards, differences in the exposure situations to which the standards apply, differences in the population groups of primary concern, and differences in the accounting of natural background radiation.

The important conclusion to be drawn from these discussions is that risks corresponding to different guidances and regulations should not be compared unless the bases of the standards and their applicability are well understood and the standards are interpreted properly. Otherwise, inappropriate and misleading conclusions about the meaning of differences in risks embodied in the standards can result.

RELATIONSHIP BETWEEN STANDARDS AND DOSES EXPERIENCED

Previous discussions in this chapter and chapter 5 have addressed the primary bases of standards (limits on levels of radionuclides in environmental media, releases to the environment, doses, or risks) in guidances and regulations for controlling radiation exposures of the public and the consistency of the standards with regard to the corresponding lifetime risks. This section considers the important question of the relationship between the standards and the doses and risks that would be experienced by exposed individuals and populations. These considerations provide important insights into the single unifying principle—namely, the ALARA objective—that is the most important in determining actual risks, irrespective of the differences in risks corresponding to the various quantitative criteria in guidances and regulations.

A discussion of the quantitative criteria in guidances and regulations that does not consider other factors that are important in controlling exposures of the public gives the impression that the criteria by themselves define acceptable risks. That impression is misleading. Irrespective of the particular environmental law under which any guidance or regulation is developed, the doses and risks experienced by exposed individuals and populations are not determined primarily by compliance with the quantitative criteria alone. This important point is illustrated in the following paragraphs.

EPA's proposed federal guidance on radiation protection of the public (EPA 1994d) incorporates the three basic principles of radiation protection set forth by ICRP (1991) and NCRP (1993a):

- *Justification* of exposures (positive net benefit).

- *Reduction* of exposures of individuals and populations to as low as reasonably achievable (ALARA), economic and social factors being taken into account, also referred to as *optimization* of exposures (ICRP 1991; 1977).

- *Limitation* of dose to individuals from all controlled sources combined.

The ALARA objective is implemented in part by establishing standards for specific sources or practices that limit doses for the exposure situations of concern to a fraction of the dose limit for all controlled sources combined, and further site-specific reductions in dose based on ALARA considerations generally are required. The important point is that the ALARA objective essentially defines a site-specific *process* for dose reduction, and the result of the process generally cannot be defined and quantified in advance in regulations.

The power of the ALARA objective in reducing doses to the public is seen by examining doses that result from operations of nuclear facilities that are regulated under the Atomic Energy Act. The average individual dose in exposed populations is only about 0.05% of the primary dose limit for the public of 1 mSv (100 mrem) per year from all controlled sources combined (NCRP 1987a); and doses to individuals who receive the highest exposures normally are no more than about 10% of the primary dose limit and often are substantially less (EPA 1989d). Therefore, for the important case of releases from operating nuclear facilities, the doses and risks experienced by most members of the public are determined largely by vigorous application of the ALARA objective, but the primary dose limit and even, in many cases, the authorized limits for specific sources or practices at a fraction of the dose limit are rather unimportant in determining actual doses and risks.

A similar example is provided by the requirements for cleanup of contaminated sites under CERCLA and its implementing regulations (40 CFR Part 300). In considering acceptable risks at contaminated sites, considerable attention normally is given to the preliminary remediation goals, including the goal for lifetime cancer risk of 10^{-4} (Luftig and Weinstock 1997; Clay 1991). However, far less attention has been given to the result that the negotiated cleanup levels at most sites, as incorporated in the ROD, correspond to risks substantially above the goal of 10^{-4} (EPA 1994b; Baes and Marland 1989). The actual cleanup levels judged to be acceptable at any site are based on a decision process that is similar to applications of the ALARA objective under the Atomic Energy Act. Therefore, for the important case of cleanup of contaminated sites, the acceptable risks at any site are determined primarily by site-specific application of the ALARA objective, not by the risk goal specified in regulations.

Another example is provided by the standards for radioactivity in drinking water (40 CFR Part 141) developed under the Safe Drinking Water Act. These standards are important because they apply to drinking-water systems used by more than half the US population and generally are being applied to protection of groundwater resources at new waste-disposal sites and at contaminated sites undergoing remediation. Although the standards specify limits (MCLs) on allowable radioactivity in community drinking-water systems, it is important to emphasize that the MCLs were based on judgments about levels of radioactivity that could be achieved, given existing levels in sources of drinking water throughout the United States and the effectiveness of available methods for water treatment, rather than an a priori judgment about acceptable risks posed by drinking water. Therefore, the MCLs are based essentially on ALARA considerations. Furthermore, the drinking-water standards are subject to change periodically on the basis of reconsideration of the costs and benefits of water treatment (EPA 1991a).

These discussions illustrate the following important points. Although guidances and regulations for controlling radiation exposures of the public contain quantitative criteria that define limits or goals for acceptable doses or risks for the exposure situations of concern, the doses and risks that would be experienced by individuals and populations are, in most cases, not determined by these criteria. For most important exposure situations, actual doses and risks that would be experienced are determined primarily by application of an ALARA process, whose outcome generally cannot be quantified in regulations. In most cases, actual limits or goals for dose or risk specified in guidances and regulations, although they represent important statements of principle and although they define an upper or lower bound on dose or risk for applying the ALARA objective, are relatively unimportant in determining actual outcomes.

Viewed in that way, all guidances and regulations for controlling radiation exposures of the public developed under any laws have as their unifying principle the objective that exposures from any source or practice should be as low as reasonably achievable (ALARA). To the extent that the ALARA objective is applied consistently in all cases and it is recognized that doses and risks that are ALARA can vary considerably depending on the particular source or practice, all guidances and regulations will be consistent with regard to doses and risks actually experienced.

SUMMARY

This chapter has reviewed EPA's existing or proposed guidances and regulations that apply to control of routine exposures of the public to naturally occurring radionuclides. No particular distinction has been made in this review between standards for naturally occurring radionuclides associated with operations of the nuclear fuel cycle, which are developed under the Atomic Energy Act, and standards for TENORM, which are developed under environmental laws other than the Atomic Energy Act and are the main concern of this study. This review has emphasized the standards that apply to naturally occurring radionuclides and the bases of the guidances or regulations.

This chapter also discussed the health risks corresponding to the quantitative criteria in the guidances and regulations that apply to naturally occurring radionuclides. The risks corresponding to the different guidances and regulations vary over several orders of magnitude, owing primarily to:

GUIDANCES AND REGULATIONS FOR NORM 155

- Differences in statutory and judicial mandates for standards, especially the difference between the traditional regulatory approach under the Atomic Energy Act, which emphasizes a limit on radiation dose and reduction in doses below the limit to as low as reasonably achievable (ALARA), and the regulatory approach under other environmental laws. These laws often emphasize a goal for risk and allowance for an increase (relaxation) in risks above the goal based primarily on technical feasibility and cost.

- Differences in the primary bases of standards, that is, the consideration that some standards are based primarily on an a priori judgment about risks that are acceptable whereas other standards are based primarily on a judgment about risks that are reasonably achievable.

- Differences in the applicability of standards, especially the considerations that some standards apply to all sources of exposure combined, whereas other standards apply only to specific sources or practices. The various standards for specific sources or practices apply to different exposure situations with different risks that are reasonably achievable.

- Differences in the population groups of primary concern in developing standards, particularly whether the standards emphasize protection of maximally exposed individuals or protection of individuals who receive the average dose in exposed populations.

- Differences in the considerations of natural background, especially whether background levels of radioactivity are important in establishing the standards.

It is important to understand those factors when judging the meaning of differences in health risks corresponding to the various guidances and regulations.

An important conclusion from the discussions in this chapter is that the large differences in health risks corresponding to the various EPA guidances and regulations do not necessarily mean that the different standards are inconsistent with regard to defining an acceptable risk to individuals or populations. Without regard for the differences in standards, as summarized above, the principle that exposures of individuals and populations should be ALARA is the most important factor in determining risks actually experienced for any controllable exposure situation. That is, largely without regard for the limits or goals specified in various guidances and regulations, application of the ALARA objective is the most important factor in determining acceptable risks. Therefore, to the extent that the ALARA objective is applied consistently to all exposure situations, all guidances and regulations would be consistent with regard to risks actually experienced, provided that it is also recognized that risks that are ALARA can vary considerably depending on the particular exposure situation.

8
Indoor-Radon Guidelines and Recommendations

National and international agencies operating under different directives have been responsible for addressing the health risk associated with indoor radon and for addressing its regulation. This chapter provides a review and comparison of national and international guidelines and recommendations regarding radon in dwellings, schools, and workplaces. It also examines the differences in the scientific information and in risk-management polices used for developing the guidelines.

In 1970, through the enactment of several statutes, the US Environmental Protection Agency (EPA) became responsible for the establishment of environmental protection standards for both radiologic hazards and chemical agents. Although there is not now a federal regulation for indoor radon, EPA has issued guidance under the Indoor Radon Abatement Act about the risk, measurement and remediation of radon in homes and schools (EPA and DHHS 1994; 1986). EPA's published guidelines and recommendations for indoor radon are different from those of other bodies that develop guidance for radiation exposure of the public.

RADON MEASUREMENT UNITS

Concentrations of radon gas in air are normally given in units of picocuries per liter (pCi/L) or becquerels per cubic meter (Bq/m^3); and 1 pCi/L is equal to 37 Bq/m^3. Concentrations of radon decay products (RDPs) normally are expressed in working levels (WL); 1 WL is defined as any combination of the short-lived RDPs in 1 L of air that results in the ultimate release of 1.3×10^5 MeV (2.1×10^{-5} J/m^3) of alpha energy—about the amount of energy emitted by the short-lived decay products in equilibrium with 100 pCi of radon. In general, equilibrium does not occur in houses, because ventilation removes some of the radon and its decay products; also, it takes time for the entering radon to produce its decay products. Because RDPs have a static charge, they plate out on walls, furniture, and other solid objects; this reduces the equilibrium ratio (ER)—the concentration of radon progeny in air divided by the concentration that would exist if the progeny were in equilibrium with the radon gas. The ER ranges between 0.3 and 0.7; an ER of 0.5 is commonly assumed as an average. At less than equilibrium, 1 WL is equal to the product of ER and the radon concentration in picocuries per liter divided by 100. A house with 150 Bq/m^3 (4 pCi/L) is likely to have 4×10^{-7} J/m^3 (0.02 WL). A working level month (WLM) is a measure of time-integrated exposure and is the product of time in working months, which is taken to be 170 hours, and working levels (WL). Thus, 1 WLM is equal to the product of average WL and hours of exposure divided by 170. Under full occupancy conditions (8,760 h/y), residence in a house at 150 Bq/m^3 results in about 1.0 WLM per year of exposure. In SI units, 1 WLM is approximately 3.5 mJh/m^3.

PATHWAYS OF HUMAN EXPOSURE

Inhalation is the principal route of radon exposure of humans. The dose contribution from the inhaled radon gas itself is small under normal conditions of exposure. Exposure to radon is due mainly to the inhalation of its short-lived decay products (polonium-214, polonium-218, lead-214, and bismuth-214), which deposit nonhomogeneously in the human respiratory tract and irradiate the bronchial epithelium. Two progeny, ^{214}Po and ^{218}Po, deliver the most important alpha-radiation dose to the lung (NCRP 1984c). About 90% of RDPs can attach initially to airborne particles (ICRP 1993b); tobacco smoke provides additional attachment sites for RDPs. The unattached fraction (10%) has a higher rate of deposition and is more efficient in delivering a dose to the critical cells (basal and secretory cells) of the lung (UNSCEAR 1993); about two-thirds of the total dose in homes from radon comes from the unattached fraction (National Research Council 1988). The assumed health effect end point of high exposure to indoor radon is radiation-induced lung cancer.

This chapter focuses on the lung-cancer risk associated with inhalation exposures to RDPs. Although risks to other tissues posed by radon can occur through ingestion of water with high radon concentrations, they are much smaller than those associated with inhalation exposure to RDPs. All other exposure pathways distribute smaller amounts of radon and progeny over a much larger tissue mass with correspondingly lower doses and risks.

HEALTH EFFECTS AND RISK EVALUATION OF RADON EXPOSURE

The existence of high mortality among miners in central Europe was recognized before 1600, and the main cause of death was identified as lung cancer in the late 19th century. It was suggested in 1924 by Ludewig and Lorenser that the cancers could be attributed to radon exposure (ICRP 1993b). EPA classifies radon as a known human carcinogen on the basis of data from epidemiologic studies of underground miners. That classification is supported by a consensus of national and international organizations (IARC 1988; National Research Council 1988; ICRP 1987b; NCRP 1984c). Further information on the deleterious health effects associated with exposure to radon has been provided by experimental studies of animals (National Research Council 1988).

The main source of quantitative information on the risks posed by radon exposure is the epidemiologic studies of miners which uses data on thousands of occupationally related lung cancers among many diverse groups of miners. The epidemiologic evidence of the induction of lung cancer after inhalation of radon comes from several cohort and case-control studies of underground miners, particularly uranium miners. The evidence has been reviewed and summarized in other reports (ICRP 1993b; National Research Council 1988; UNSCEAR 1988; UNSCEAR 1986). Most of the data are consistent with the assumption of a proportional relationship between risk and cumulative exposure (linear, no-threshold response model). The exception to linearity occurs at very high exposures (over 2,000 WLM), where the response per unit exposure decreases; this exception is attributed partly to the reduced life expectancy of the miners at such high exposures (ICRP 1987b).

The epidemiologic findings have to be extrapolated to provide risk estimates for long periods of exposure and for populations other than those studied. For estimating lifetime risk from data covering shorter periods, projection models are used. Different types of risk-projection models have been proposed to estimate the possible lifetime risk of lung cancer posed by inhaled radon progeny in homes on the basis of the results of the epidemiologic studies of miners (National Research Council 1988; ICRP 1987b; NCRP 1984c).

The National Council on Radiation Protection and Measurements (NCRP) model is an attributable-risk projection model based on information obtained from several groups of underground miners in the United States, Canada, and central Europe (NCRP 1984c). The model expresses lung-cancer risk uniformly with time after exposure, with the restriction that tumors do not occur either before a 5-y latent interval or before the age of 40. The model uses an initial, age-averaged risk coefficient as derived from data on miners and assumes a decrease in the initial potential excess rate with time after exposure according to an exponential function. That functional structure provided an age dependence so that the model would fit the observations of lung-cancer frequency among radon-exposed miners. The NCRP risk reduction with time since exposure is supported by the followup studies of underground miners (National Research Council 1988).

The International Commission on Radiological Protection (ICRP) model is a constant relative-risk projection model based on the lung-cancer incidence data from the uranium miner cohort studies (United States, Canada, and Czechoslovakia) and on information from the atomic-bomb survivors (ICRP 1987b). The model assumes that the excess risk of lung cancer in miners associated with a given radon exposure is constant with age and over time after the end of exposures. ICRP made three modifications to the radon relative risk coefficients from the miner data to reflect presumed differences in residential indoor exposures. First, because of potential cocarcinogenic influences that might be present in the mines but not indoors (such as exposures to diesel fumes, dust, and other forms of radiation), ICRP assumed that the risk coefficients for residential indoor exposures would be 80% of those for mine exposures. Second, because of potential differences in breathing rate and the unattached fraction between residential and mine exposure conditions, ICRP assumed that the observed dose of alpha radiation per unit of cumulative radon exposure for the general population is only 80% of that for miners. Third, on the basis of findings from studies of the atomic-bomb survivors, ICRP assumed a risk coefficient for exposure of people under 20 y old that was 3 times the risk coefficient for people 20 or older; this had the effect of increasing the overall lifetime risk by about 40%. The latest recommendations by ICRP (1993b) retain the multiplicative risk-projection model as in previous publications (ICRP 1987b).

The BEIR IV committee model (National Research Council 1988) is a relative-risk projection model based on reanalyses of cohort studies of underground miners (US and Canadian uranium miners and Swedish iron miners). The model assumes that the rate of excess lung cancer due to radon exposures increases with age-specific baseline lung-cancer mortality. The BEIR IV modified relative-risk model is somewhat different from the ICRP 50 model (ICRP 1987b) in the added assumptions about the effects of time since exposure and attained age. It incorporates the BEIR IV finding that excess relative risk in the miners decreased with time since exposure and attained age. Direct evidence on the sensitivity of children to radon is sparse. The BEIR IV committee did not find an effect of age at first exposure after controlling for other correlates with age (National Research Council 1988). That is consistent with the publication of the BEIR V report, which found no evidence of dependence of lung-cancer risk on age at exposure for external radiation (National Research Council 1990). The effect of any higher relative risk in the period soon after exposure of children would probably be offset by the decrease in excess relative risk with time.

Although each of the above three models incorporates risk coefficients derived from the studies of miners, the biologic assumptions underlying the models differ. The different features of these risk-projection models are summarized in table 8.1 (National Research Council 1991). The NCRP model (NCRP 1984c) assumes additivity of the risks posed by radon progeny and the background risk of lung cancer and a time-dependent decline in risk after exposure. In contrast, the ICRP model (ICRP 1987b) assumes that the background rate is multiplied by the additional risk associated with radon progeny. The model developed by the National Research Council (National Research Council 1988) is also multiplicative, but it incorporates a time-dependent decline in risk after exposure. With regard to the relationship between exposure in the mining environment and exposure in the home environment, the three models make different assumptions. The BEIR IV model makes no adjustment, whereas the ICRP model reduces the risk by 20% for adults in the general population, and the NCRP model increases the risk by 40% for the residential exposures, because of a higher calculated unattached fraction. In addition, the ICRP model increases risk for exposures before the age of 20 y, and the NCRP model assumes that risk commences at the age of 40 y. In the BEIR IV model, risk varies with attained age. With regard to smoking, the NCRP model is additive, whereas the other two models are multiplicative (National Research Council 1991).

The estimate of risk based on chronic occupational exposure to radon in the BEIR IV report (National Research Council 1988), given as a lifetime fatality coefficient, is 3.5×10^{-4} per WLM for a US population. The corresponding ICRP value is 3×10^{-4} per WLM based on a "reference" population with somewhat lower baseline cancer mortality. EPA's estimates of lung-cancer risk posed by radon exposure at 150 Bq/m^3 (4 pCi/L) —1.6×10^{-3} for never-smokers and 3×10^{-2} for smokers—are based on the report of the National Research Council (1988) and an adjustment recommended by the National Research Council (1991). EPA has made two adjustments to the BEIR IV model in estimating radon risks. In the first, age-specific baseline lung-cancer mortality was adjusted by eliminating projected deaths due to an average background radon exposure of 0.24 WLM per year, reducing the lifetime risk estimates by about 10%. The second was based on differences in dose to the bronchial epithelium per unit of radon-progeny exposure in mines and homes due to a number of physical and biologic factors that are expected to differ in the two environments. Among the factors considered in the 1991 National Research Council report are age, sex, aerosol size distribution, unattached fraction of radon progeny, breathing rate and route (oral vs. nasal), pattern and efficiency of deposition of radon progeny, solubility of radon progeny in mucus, and growth of aerosols in the respiratory tract. This comparison of exposure-dose relations in the mining and home environments indicated that the dose per

Table 8.1 Comparison of principal risk-projection models for radon and lung-cancer

Feature	NCRP	ICRP	BEIR IV
Form of model	Attributable-risk	Relative-risk	Relative-risk
Time-dependent	Yes; risk declines exponentially after exposure	No	Yes; risk declines as time since exposure lengthens
Lag	5 y	10 y	5 y
Effect of age at exposure	No effect of age at exposure	Threefold increased risk for exposures before age of 20 y	No effect of age at exposure
Age at risk	Risk commences at age of 40 y	Relative risk does not change with age	Lower risk at age of 55 y and older
Dosimetry adjustment	Increased risk for indoor exposure	Decreased risk for indoor exposure	No adjustment
Risk coefficient	10×10^{-6}/y per WLM	Excess relative risks: 1.9%/WLM at ages 0-20 years and 0.64%/WLM for ages 21 years and above	Excess relative risk of 2.5%/WLM but modified by time since exposure

Source: National Research Council (1991).

unit of exposure to radon progeny is about 30% lower in the home environment. Therefore, in calculating the risks associated with residential exposures, EPA multiplied the risk coefficient in the BEIR IV model by a factor of 0.7 (EPA 1992c).

The assumption underlying the EPA estimates of radon risk by smoking category is that radon risk varies in proportion to smoking risk (radon and smoking act multiplicatively in causing lung cancer). The data source used by EPA for the prevalence of and relative risks associated with smoking was the surgeon general's report (DHHS 1989). The calculated lung-cancer death rates in each smoking category in the general population were compared with the average lung-cancer death rate for the general population to obtain radon risk multipliers. For example, the number of lung-cancer deaths per 100,000 current smokers in the general population (males and females combined) is 10,329. That is 2.33 times the 4,433 lung-cancer deaths expected in the general population, averaged over all smoking categories (EPA 1992c). From the presumed multiplicative interaction between radon and smoking, the radon risk among current smokers also would be about 2.33 times the radon risk for the general population. The risk multipliers were used in conjunction with a standard life-table analysis based on 1980 vital statistics and the EPA-adjusted BEIR IV relative-risk model to estimate the lung-cancer risks. As discussed previously, the risk coefficients used in the BEIR IV risk model were adjusted by a factor of 0.7 to correct for an estimated lower bronchial radon dose per WLM in homes than in mines (National Research Council 1991). The lung-cancer baseline risk was also adjusted for an annual background radon exposure of 0.24 WLM. Table 8.2 shows the risks for never-smokers and current smokers, with the risks for the general population, for selected radon exposures (EPA 1992c). The lung-cancer risk to current smokers associated with exposure to radon progeny is substantially greater than the radon risk to never-smokers.

The analysis of the effects of smoking on Rn risk is subject to uncertainty about the nature of the interaction (multiplicative or submultiplicative), the variation in the relative risk associated with smoking by age and gender, the changes in age-specific relative risks for smokers as smoking habits and types of cigarettes change over time, and the influence of environmental and passive cigarette smoke on Rn risk (EPA 1992c).

More quantitative information on lung-cancer risks posed by exposure to radon progeny was provided by a joint analysis of data from 11 studies of underground miners (Lubin and others 1994). The authors examined over 2,700 lung-cancer deaths that occurred among 68,000 miners. The analyses confirm the linear relationship between cumulative exposure to radon progeny and lung-cancer risk and a decrease in excess relative risk (ERR) per WLM with attained

Table 8.2 Estimated lifetime lung-cancer risk for never-smokers, current smokers, and the general population [a]

Radon Level			Lifetime Lung-cancer Risk (per person)		
Bq/m^3	pCi/L	WLM/y	Never-smokers	Current Smokers	General Population [a]
740	20	3.9	7.7×10^{-3}	1.4×10^{-1}	6.2×10^{-2}
370	10	1.9	3.9×10^{-3}	7.1×10^{-2}	3.1×10^{-2}
300	8	1.6	3.1×10^{-3}	5.7×10^{-2}	2.5×10^{-2}
150	4	0.77	1.6×10^{-3}	2.9×10^{-2}	1.3×10^{-2}
74	2	0.39	7.8×10^{-4}	1.5×10^{-2}	6.4×10^{-3}
46	1.3	0.24	4.9×10^{-4}	9.3×10^{-3}	4.0×10^{-3}
15	0.40	0.077	1.6×10^{-4}	3.0×10^{-3}	1.3×10^{-3}

[a] Risk estimates of EPA (1992c).

age, with time since exposure, and with time after cessation of exposure. Among miners first exposed to radon progeny under the ages of 10-20 y, the ERR per WLM was not related to age at first exposure. The report also noted that for equal total exposure, exposures of long duration and low rate (typical of exposures in homes) were more harmful than exposures of short duration and high rate. Among cohorts with tobacco-use information, the slope of the radon exposure-response function for never-smokers was 3 times that for smokers, indicating a much greater risk for never-smokers relative to their background risk of lung cancer from all causes. Assuming that the miner-based findings apply to residential radon exposure, the study estimated that about 9% of all lung-cancer deaths among residents of single-family dwellings in the United States could be attributable to indoor radon exposure. The estimates are similar to estimates based on the BEIR IV risk model. On the basis of the relative differences in ERR per WLM for smokers and never-smokers, it was estimated that indoor radon-progeny exposure could be responsible for 10-12% of the lung-cancer deaths among smokers and 28-31% of the lung-cancer deaths among never-smokers. For the roughly 15,000 estimated lung-cancer deaths in the United States in 1993 that might be attributable to indoor radon-progeny exposure, those percentages translate to about 10,000 lung-cancer deaths among smokers and 5,000 among never-smokers (Lubin and others 1994).

Several studies have aimed at detecting the correlation between the incidence of lung cancer and exposure to radon in dwellings. The results are mixed. A study of radon and lung cancer in women, with 480 lung-cancer cases and 442 controls, has reported a statistically significant trend with increasing residential radon concentration after adjusting for smoking and age (Schoenberg and others 1990). However, another study showed no statistically significant association between radon exposure in homes and lung-cancer risk (Blot and others 1990). A study of indoor radon and lung cancer in Swedish women, with 210 lung-cancer cases and an equal number of controls, reported increasing trends of lung-cancer risk with radon exposures exceeding 150 Bq/m^3 (Pershagen and others 1992).

Recently, Lubin and Boice (1997) provided additional information on the risk of lung cancer associated with indoor radon. They conducted a meta-analysis of eight residential case-control studies that included at least 200 case subjects each and that use long-term indoor radon measurements. The analysis included a total of 4,263 lung-cancer cases and 6,612 control subjects. From the published results of each study, relative-risk (RR) estimates for various categories of radon concentration were obtained, and weighted linear-regression analyses were performed. The combined trend in RR was significantly different from zero, and an estimated RR of 1.14 (95% CI, 1.0-1.3) at 150 Bq/m^3 was found. The exposure-response trend was similar to model-based extrapolation found. The exposure-response trend was similar to model-based extrapolation from miners and to RRs computed directly from miners with low cumulative exposures.

A summary of lifetime risk estimates of lung-cancer mortality associated with chronic exposure to radon progeny estimated by various organizations is provided in table 8.3.

OVERVIEW OF RADON GUIDELINES AND RECOMMENDATIONS FOR DWELLINGS

National Council on Radiation Protection and Measurements (NCRP)

One of the earliest recommendations for domestic radon exposure in the United States was developed by NCRP on the basis of available data on lung-cancer risk (NCRP 1984c). The NCRP recommendation states that an excess risk of death from lung cancer of 2% (a doubling of the average background risk of lung cancer) or more over a lifetime for individuals exposed to enhanced levels of radon decay products should be avoided. The NCRP recommendation was based on evaluation of the lung-cancer risk and the avoidance of an unacceptable exposure and thus risk. The recommendation may be considered as an "upper bound based on maximum tolerable risk." The excess risk of 2% corresponds to an annual exposure of 2 WLM (equivalent to 8-10 pCi/L if 0.4 or 0.5 is used for the equilibrium ratio) and is the recommended NCRP remedial action level for radon exposure. It is about 10 times the average background exposure of 0.2 WLM assumed for the US population. NCRP recommended that exposure above the remedial action level be reduced using appropriate actions. It also stated that exposures just below the remedial action level might not be acceptable to some individuals, who could of course try to reduce their exposures further.

The assumption by NCRP of an excess risk of death from lung cancer of 2% posed by a lifetime exposure of 2 WLM was based on the available underground-miner epidemiologic data at the time and on the attributable-risk projection model (NCRP 1984b). The NCRP estimate is less than the risk estimated by more-recent projection models, such as those of BEIR IV (National Research Council 1988) and ICRP 50 (1987b). The most recent recommendations of NCRP (1993a) retain the same action level for indoor radon as previously recommended by NCRP (2 WLM) on the basis of an excess lifetime risk of no more than 10 times the risk associated with the average annual background levels found in homes and consideration of the feasibility of remediation.

Table 8.3 Comparison of lifetime-risk coefficient estimates associated with chronic exposure to radon progeny

Reference	Excess lifetime lung cancers: Deaths per 10^6 persons per WLM
ICRP (1993b)	300
EPA (1992c)	220[a]
EPA (1989b)	305[b]
EPA (1989b)	360[c]
National Research Council (BEIR IV) (1988)	350
EPA (1987b)	460[d]
ICRP (1987b)	230
NCRP (1984c)	130

[a]Based on BEIR IV model adjusted by subtracting projected deaths due to average background radon exposure of 0.24 WLM/y and adjusting risk factor from occupational exposure to general public with a factor of 0.7.
[b]Based on BEIR IV model adjusted by eliminating deaths due to average background radon exposure of 0.25 WLM/y.
[c]Average of BEIR IV and ICRP 50 models.
[d]Based on constant relative-risk model.

US Environmental Protection Agency

The discovery of extremely high levels of indoor radon in the northeastern United States in 1984 created a major public-health concern (Oge 1992). As a result, EPA has initiated activities to increase the public's awareness of radon risk and understanding of the options for reducing exposures. EPA used the data and knowledge available at the time to begin immediately reducing the risks posed by elevated radon levels. EPA adopted a nonregulatory approach to the problem and recommended a primary action level of 150 Bq/m^3 (4 pCi/L) as the point above which mitigation is always advised to reduce radon in existing buildings (EPA and DHHS 1986).

The 1986 EPA and DHHS recommendation stems from the guideline originally developed for homes built on uranium mill tailings in Grand Junction, CO (Harley 1996). The guideline, published in 1976, was developed by scientists in the Colorado Department of Health, the Surgeon General, Public Health Service, and the Atomic Energy Commission. The guideline was as follows:

- At 0.05 WL in excess of background, remedial action is indicated (this is equivalent to 10-12 pCi/L, depending on whether 0.4 or 0.5 is chosen for the equilibrium ratio of the radon decay products).

- From 0.01 to 0.05 WL in excess of background, remedial action may be suggested.

- At less than 0.01 WL in excess of background, no action is indicated.

In Colorado, the Department of Energy (DOE) was attempting to remediate homes built on tailings to the guideline of "no action necessary" levels of less than 0.01 WL above background. The background level of indoor radon decay products is about 0.01 WL, so DOE was attempting to remediate to a gross value (contamination plus background) of 0.02 WL. The EPA radon action level of 4 pCi/L (equivalent to 0.02 WL at 50% equilibrium of the radon decay products) is consistent with the DOE undertakings to remediate to the "no action necessary" level of the original guidelines (Harley 1996).

In its revised guidance, EPA has maintained the same 4-pCi/L action level for indoor radon (EPA and DHHS 1994). EPA sought to balance several factors, including the findings of its technical analysis of risk, cost-effectiveness, and the practical limitation of radon testing accuracy and mitigation technology. It examined five action levels (2, 3, 4, 8, and 20 pCi/L) for the guideline. Higher action levels did not reduce the population risk posed by exposure to radon nearly as much as did lower action levels. The agency has focused its attention on action levels of 150 Bq/m^3 (4 pCi/L) or lower. An action level of 4 pCi/L was determined to be incrementally cost-effective (EPA 1992c). For example, the average cost per life saved by using this action level is about $700,000—well within the range of the costs per life saved by other government programs and regulations, such as highway safety, air-transportation safety, and occupational safety. Furthermore, EPA believes that the 150-Bq/m^3 (4-pCi/L) action level is technologically achievable in the vast majority of homes. The current guideline also recommends that mitigation be considered for indoor radon concentrations in the range of 75-150 Bq/m^3 (2-4 pCi/L), provided that concentrations can be reduced to less than 75-150 Bq/m^3. EPA has emphasized several priorities regarding indoor radon: target the highest-risk areas first, promote radon-resistant new construction, support testing and mitigation in connection with real-estate transactions, use information and motivation programs to promote public awareness, and develop a coordinated research plan for radon-related issues.

As previously discussed, EPA has relied primarily on relative-risk projection models to estimate radon risks to the public (National Research Council 1988). EPA's Science Advisory Board (SAB) did not recommend use of the NCRP Report 78 model (NCRP 1984b), which is an absolute-risk model. Absolute-risk models have been described as less appropriate for the estimation of lifetime radon risk because they do not assume the temporal correlation with the baseline lung-cancer rate as indicated by available data (ICRP 1987b). Moreover, the NCRP model presumes that the effects of radon and cigarette smoking are additive, contrary to epidemiologic evidence of a near-multiplicative relationship (EPA 1992c). The SAB also recommended, on the basis of two pieces of recent information, that the agency use only the BEIR IV model and discontinue use of the ICRP 50 model. The first was evidence from epidemiologic studies of a decrease in lung-cancer risk with time since exposure, which had been incorporated into the BEIR IV model but not the ICRP 50 model. The second was the publication of the BEIR V report (National Research Council 1990), which found no evidence of dependence of lung cancer on age at exposure for external radiation.

For exposure at the 4 pCi/L level, the EPA's estimated lifetime risks of fatal lung cancer are 1.6×10^{-3} for never-smokers and 3×10^{-2} for smokers (EPA and DHHS 1994). EPA's risk estimates at 4 pCi/L are in line with the recommendations of ICRP (1991; 1987b) and BEIR IV (National Research Council 1988). Using the adjusted BEIR IV risk estimates with results from the National Residential Radon Survey (EPA 1992b), EPA calculated about 14,000 lung-cancer deaths per year caused by indoor radon exposure of the US population, with an uncertainty range of 7,000-30,000.

The following sources of uncertainty in the estimate of lifetime risk were addressed by EPA: statistical variability in the miner data, projection of risk beyond the period of epidemiologic followup (projection of risk over time), age dependence of risk, extrapolation from mines to homes, the influence of mine exposures other than to radon, the exposure-rate effect, extrapolation to females, and the relationship between radon risk and smoking risk (EPA 1992c).

International Commission on Radiological Protection

ICRP is one of the principal sources of guidance for radiation-protection policies for most international bodies. Its recommendations for protection of the public from radiation distinguish two circumstances of exposure: one in which human activities introduce new sources or modes of exposure and thus increase the overall exposure, and one in which they decrease the exposure to existing sources. The first it calls practices and the second, interventions (ICRP 1991). Reducing exposure due to pre-existing natural sources, such as indoor radon in existing structures, is clearly an intervention. The system of intervention is based on two principles. The first, justification of intervention, requires that an intervention itself do more good than harm; the reduction in the radiation detriment should be enough to justify the harm, including the cost of intervention. Under the second principle, optimization of intervention, the form, scale, and duration of the intervention are chosen to obtain the maximum net benefit.

ICRP (1991) recognized the complex problems involved in controlling exposure to indoor radon, which is by far the largest source of average human exposure to natural background radiation. It recommended that the choice of an action level for radon should depend not only on the magnitude of risk to individuals, but also on the likely scale of action required and its economic implications for communities and individual homeowners. ICRP also stated that the particular national action level chosen could be one that defines a sizable, but not unmanageable, number of houses in need of remedial work. Intervention is to be applied to reduce the risk to those most highly exposed and not primarily to reduce the collective dose to the population.

Because of its widespread occurrence and relatively high concentrations, radon has been treated separately by ICRP. ICRP recommends that countries carry out radon surveys to identify radon-prone areas, defined as those in which more than 1% of homes contain radon at more than 10 times the national average. That is a way to identify a manageable number of homes with the greatest risk to occupants (ICRP 1993b).

The principle of justification is used to set action levels at which intervention would almost always be justified to reduce radon exposure. ICRP considered that simple countermeasures would be virtually certain to be justified to avert an effective dose of 10 mSv in a year (the upper bound of the maximum tolerable dose to individuals).

The principle of optimization will lead to a lower action level than 10 mSv, but not lower by as much as a factor of 10, because that would often correspond to the national average radon exposure. Using an effective dose per unit exposure of 4 mSv per WLM (ICRP 1993b) for exposure in homes, ICRP recommended a radon concentration of 600 Bq/m^3 as that at which action is almost certain to be justified, and it expected optimization to suggest an action level no lower than 200 Bq/m^3. It recommends that national authorities set an action level of about that and advise residents whose homes are at higher levels that they should initiate remedial measures. It recommends that new homes be designed to avoid the problem of high radon. ICRP no longer sets another, lower target level for new homes as it did previously (ICRP 1984).

ICRP and EPA use similar risk-management approaches to address control of the hazard posed by indoor radon (Overy and Richardson 1995). They agree that risks posed by domestic radon should not be treated in the same manner as other risks, because the social and economic effects are greater and more complex.

International Atomic Energy Agency

The International Atomic Energy Agency (IAEA) basic safety standards for protection against ionizing radiation have been recently published (IAEA 1996a). The guidelines specific to indoor radon are based on the ICRP recommendations (ICRP 1993b; 1991). The main IAEA recommendation is that the optimized action level related to chronic exposure involving radon in dwellings should fall within a yearly average ^{222}Rn concentration of 200-600 Bq/m^3 in air. For the optimized action levels, account should be taken of the benefits and costs assessed in the remedial action plan.

Commission of the European Communities

The current indoor-radon control policy recommendations of the Commission of the European Communities (CEC) were published in the *Official Journal of the European Communities (EC)* in 1990 (EC 1990). The recommendations are not legally binding, but constitute, within the EC, the reference framework for the initiation of policies at the national level. The main CEC recommendations, adopted mostly from ICRP (1984), are:

- To set a reference level of 400 Bq/m^3, above which consideration should be given to reducing radon concentrations in existing homes, and a design level of 200 Bq/m^3 for all new dwellings.

- To use annual average radon concentrations as a basis for radiologic protection decisions and to develop criteria for identifying regions, sites, and building characteristics likely to be associated with high indoor radon concentrations.

- To have national authorities provide information to the public on radon exposure, risk, and available remedial measures.

Other

Various countries' and organizations' current recommendations for action levels for existing houses and for upper limits (bounds) in new buildings are summarized in table 8.4. Additional comments and information are also provided in the tabular summary.

Values for existing dwellings are mostly of an advisory nature. In Sweden and Switzerland, the levels are legally enforced, and both apply "recommended" action levels, lower than the regulatory limits, above which remediation is advised. Although most guidances are not enforced standards for limiting indoor radon exposures of the public, they are widely used as de facto standards in the real-estate and insurance industries. For example, in the United States, lending institutions often require radon concentrations (based on short-term measurements) less than the EPA action level of 150 Bq/m^3 (4 pCi/L) as a condition for financing home purchases.

Of the 15 member states of the European Union, only Austria and Finland have adopted the values proposed in EC (1990). Belgium, Germany, Ireland, Luxembourg, Sweden, and the United Kingdom have adopted somewhat different values. In the Netherlands, an approach based on limiting individual risk has been adopted. Peak radon concentrations in the Netherlands are relatively low in comparison with those of other countries. The 20-Bq/m^3 reference level in the Netherlands reflects the low indoor radon concentrations generally found and provides lifetime risks of 10^{-4} or less.

Australia, Austria, Germany, Ireland, Norway, Switzerland, and the United Kingdom have set their action levels on the basis of recommendations of ICRP (1993b) and other considerations regarding cost, risk, and feasibility. In Canada, cost-benefit analysis was used as a basis for a radon reference level of 800 Bq/m^3. A similar approach was initially adopted in Sweden, which has recently reduced its reference level after the publication of ICRP 65 (1993b).

Table 8.4 Summary of national and international recommendations for indoor radon in dwellings

Country/Organization (Reference)	Radon concentration (Bq/m^3)[a]		Year Set	Additional Information
	Reference (Action) level, Existing Buildings	Upper Bound, New Buildings		
Australia (Colgan and Gutierrez 1996; Ahmed 1992)	200	200	1990	Based on basic safety standards of radiation protection of IAEA (1982)
Austria (Colgan and Gutierrez 1996)	400	200	1992	Adopted CEC values
Belgium (Colgan and Gutierrez 1996)	400	400	1995	
Canada (Colgan and Gutierrez 1996; Ahmed 1992)	800	800	1988	Policy and risk management decision Specify new construction techniques in uranium-mining areas Long-term objectives: outdoor level
China (Colgan and Gutierrez 1996)	200	100	---	Values under consideration

Table 8.4 (continued)

Country/Organization (Reference)	Radon concentration (Bq/m³)[a]		Year Set	Additional Information
	Reference (Action) level, Existing Buildings	Upper Bound, New Buildings		
Finland (Colgan and Gutierrez 1996)	400	200	1992	Adopted CEC values Radon measurement protocol: at least 2 mo duration New building practices specified for radon-prone areas Remediation funded by gov't. up to 20% of total costs
France (Colgan and Gutierrez 1996; Ahmed 1992)	400	200	---	CEC values applicable
Germany (Colgan and Gutierrez 1996; Ahmed 1992)	250	250	1988	Government offers free on-demand radon measurements and grants financial help for mitigation
Greece (Colgan and Gutierrez 1996)	400	200	---	CEC values applicable
India (Ahmed 1992)	150	---	---	US EPA guidelines applicable
Ireland (Colgan and Gutierrez 1996; Ahmed 1992)	200	200	1991	Reduction measures recommended for new construction in radon-prone areas

Table 8.4 (continued)

Country/Organization (Reference)	Radon concentration (Bq/m³)[a]		Year Set	Additional Information
	Reference (Action) level, Existing Buildings	Upper Bound, New Buildings		
Italy (Colgan and Gutierrez 1996)	400	200	---	CEC values applicable
Japan (Ahmed 1992)	None	none	---	Radon policy will be set after national survey is completed
Luxembourg (Colgan and Gutierrez 1996; Ahmed 1992)	250 150	250 150	1988 1992	Government provides radon measurement to homeowners free of charge
Netherlands (van Brederode and Bosnjakovic 1992)	20	20	1994	Based on maximum allowable lifetime risk level of 10^{-4} National average in dwellings is low: 29 Bq/m³
Norway (Colgan and Gutierrez 1996; Ahmed 1992)	200	<60-70	1990	The 60-70 Bq/m³ is national average range. Remediation advised for radon levels of 200-800 Bq/m³ if cost is less than ~$400
Portugal (Colgan and Gutierrez 1996)	400	200	---	CEC values applicable
Spain (Colgan and Gutierrez 1996)	400	200	---	CEC values applicable

Table 8.4 (continued)

Country/Organization (Reference)	Radon concentration (Bq/m³)ᵃ		Year Set	Additional Information
	Reference (Action) level, Existing Buildings	Upper Bound, New Buildings		
Sweden (Colgan and Gutierrez 1996; Ahmed 1992)	200		1994	Recommended upper limit for remediation
				Radon measurements made only during heating season for 1-3 mo
				Recommended reduction to 70 Bq/m³ if simple measures are possible
				Radon-reduction measures must be included in new construction in areas identified by "risk maps"
				Government covers 50% of mitigation costs when radon exceeds 400 Bq/m³ up to about $2,500
	400	200	1994	Regulatory limits
Switzerland (Colgan and Gutierrez 1996)	1000	400	1994	Regulatory limits
	400			Recommended upper limit for remediation

GUIDELINES FOR EXPOSURE TO TENORM

Table 8.4 (continued)

Country/Organization (Reference)	Radon concentration (Bq/m³)[a]		Year Set	Additional Information
	Reference (Action) level, Existing Buildings	Upper Bound, New Buildings		
UK (Jones and Dron 1992)	400 200	400 200	1987 1990	Government offers free on-demand radon measurements to public Radon measurements protocol: at least 3 mo Reduction measures must be included for new construction in radon-prone area Financial assistance to needy toward remediation
US (EPA and DHHS 1994; EPA 1992; EPA and DHHS 1986; NCRP 1984)	150 (4 pCi/L)	150	1994	EPA guideline long-term objectives: outdoor level Some states or local areas may require radon-resistant construction features Mitigation is advised within range of 75-150 Bq/m³
	370		1984	NCRP recommendation

Table 8.4 (continued)

Country/Organization (Reference)	Radon concentration (Bq/m³)[a]		Year Set	Additional Information
	Reference (Action) level, Existing Buildings	Upper Bound, New Buildings		
CEC (Janssens 1992; Venuti and Piermattei 1992; EC 1990)	400	200	1990	Based on ICRP (1991) recommendations Member countries could set different reference levels depending on their own balance between health risks and social-economic impact of remedial action
ICRP (1993b; 1984)	400	200	1984	Recommendations for dwellings
	200-600	200-600	1993	Recommendations for dwellings
IAEA (1996a)	200-600	200-600	1994	Based on ICRP (1993; 1991)
Nordic Countries (Denmark, Finland, Iceland, Norway and Sweden) (Nordic Radiation Protection Institutes 1986)	400	100	1986	Recommend that remediation should be considered even at 100 Bq/m³

[a] 1 Bq/m³ = 0.027 pCi/L

Similarly, the reference level in Luxembourg was reduced from 250 to 150 Bq/m^3 in 1992. The value of 150 Bq/m^3 recommended by the US EPA is based as much on technologic limitations and cost-benefit analyses as on health risks (EPA 1992c).

Although national and international guidance for radon in dwellings varies, most values have similar scientific and technical bases and are within only a factor of about 2 from each other. Most differences are related to policies and risk-management decisions by the various bodies that develop radon guidance.

RADON ACTION LEVELS FOR WORKPLACES

Workplaces are defined as ordinary places of work, such as offices, schools, stores, theaters, libraries, and hospitals (Clarke 1995). For the purposes of this discussion, workplaces do not include nuclear fuel-cycle facilities. Radon is present in all workplaces. In some cases, such as uranium mines, exposure to radon is already subject to occupational control. Because of the lower occupancy rate and associated lower accumulated exposure, radon in ordinary workplaces is widely ignored or of secondary concern compared with radon in dwellings. Some countries, however, have placed strong emphasis on radon measurement and remediation in schools (for example, Sweden, the United Kingdom, and the United States).

Exposure to radon in schools is often viewed separately from that in other workplaces, to emphasize protection of young people and because large numbers of people are potentially exposed. Although occupancy rates of schools are lower than those of dwellings, some countries have adopted the same action levels for both. In Switzerland, the reference level of 400 Bq/m^3 for schools is lower than the legally enforced 1,000 Bq/m^3 limit for dwellings but is equal to the level above which remediation of homes is recommended. In the United Kingdom, Finland, Switzerland, and Norway, the values for schools are legally enforced. The US EPA advisory reference level of 150 Bq/m^3 for schools is the same as for dwellings.

There is considerable diversity in the approach to radon in workplaces, and the range of action levels is wider than that for dwellings (table 8.5). Levels range from a target value of 20 Bq/m^3 in the Netherlands to a statutory limit of 3,000 Bq/m^3 in Switzerland.

ICRP (1993b) recommends that intervention levels for exposure to radon in homes be carried over to workplaces for exposure to radon. The level at which intervention in the workplace is almost certainly justified is the same as

Table 8.5 National and international reference (action) levels for radon in workplaces excluding those linked to the nuclear fuel cycle[a]

Country or Organization	Radon Concentration (Bq/m^3)[b]		Status
	Aboveground Workplaces	Schools	
Australia	1,000	1,000	Advisory based on IAEA recommendations
Austria	400	400	Existing advisory
	200	200	New advisory
Canada	None	800	Advisory
Finland	400	400	Legally enforced
Germany	None	250	Advisory
Ireland	None	150	Advisory
Luxembourg	150	150	Advisory
Netherlands	20	20	Advisory
Norway	800	800	Legally enforced
Sweden	400	400	Existing advisory
	400	400	Legally enforced, new
Switzerland	3,000	400	Legally enforced
United Kingdom	0.01 WL in any 8-h period		Legally enforced
US EPA	None	150	Advisory
IAEA	1,000	1,000	Advisory
ICRP	500-1,500	500-1,500	Advisory

[a]Colgan and Gutierrez (1996).
[b]1 Bq/m^3 = 0.027 pCi/L

in homes—10 mSv in a year. Because of the different occupancy factor—2,000 h at work and 7,000 h at home each year—and an effective dose per unit exposure of 5 mSv per WLM (ICRP 1993b), one arrives at a radon concentration in the workplace of about 1,500 Bq/m^3 as the level at which action is almost certainly justified. With optimization, the suggested range within which an action level should be set is 500-1,500 Bq/m^3. The IAEA remedial action level for chronic exposure involving radon in workplaces is a yearly average ^{222}Rn concentration of 1,000 Bq/m^3 (IAEA 1996a). This guideline appears to be based on the average range of 500-1,500 Bq/m^3 recommended by ICRP (1993b).

SUMMARY AND CONCLUSIONS

- The first recommendations for domestic radon exposure in the United States were developed by NCRP in 1984. The NCRP recommendations were based on evaluation of the lung-cancer risk and the avoidance of an unacceptable risk. A personal avoidance of lifetime lung-cancer risk of 2% was proposed, that is, avoidance of a continuous exposure to 2 WLM (equal to radon concentration of about 10 pCi/L). This lifetime risk was compared with a normal lung-cancer risk in smokers of about 10% and in nonsmokers of about 1%.

- On the basis of different risk projection models, the BEIR IV Committee and other major radiation protection organizations (ICRP) have estimated that higher lung-cancer risks are associated with indoor radon exposure since the NCRP publication of 1984. Although all three models (NCRP, BEIR IV, and ICRP) incorporate risk coefficients derived from the studies of miners, the biologic assumptions underlying the models differ.

- EPA's current estimates of lung-cancer risk associated with indoor radon exposures are based on the BEIR IV report and later adjustments. EPA risk estimates related to domestic radon exposure are 1.6×10^{-3} for never-smokers and 3×10^{-2} for smokers at 1 WLM (equal to 4 pCi/L, which is recommended as the remedial action level). EPA guidelines are generally comparable with the recommendations of ICRP.

- The above approaches for estimating lung-cancer risks from the miner data require numerous adjustments for estimating comparable risks in homes, and those approaches assume that cancer incidence observed in miners at high radon levels can be extrapolated linearly to zero exposure. Other sources of uncertainty include statistical variability in the miner data, the age dependence of risk, extrapolation to females, the relationship between radon risk and smoking risk, and the impact of risk extrapolation of the different levels and types of particles in uranium mines and in homes.

9
Other Guidances for TENORM

This chapter reviews guidances for TENORM other than indoor radon that have been developed by organizations other than the Environmental Protection Agency (EPA). Guidances for TENORM developed by EPA, including the guidance for indoor radon, are discussed in chapter 7, and guidances for indoor radon developed by EPA and other organizations are discussed in chapter 8.

This chapter discusses guidances that are directly applicable, or potentially relevant, to TENORM other than indoor radon that have been developed by the National Council on Radiation Protection and Measurements, the International Commission on Radiological Protection, the Health Physics Society, the Nuclear Regulatory Commission, the Department of Energy, various state governments and the Conference of Radiation Control Program Directors, the International Atomic Energy Agency, and the Commission of the European Communities and other nations. The roles of these organizations in radiation protection of the public and the regulation of TENORM are discussed in chapter 6. This chapter also discusses the important issue of the transferability of standards from one exposure situation to another.

NATIONAL COUNCIL ON RADIATION PROTECTION AND MEASUREMENTS

National Council on Radiation Protection and Measurements (NCRP) Report No. 116 (NCRP 1993a) contains NCRP's most recent recommendations on limitation of exposure of workers and the public to ionizing radiation. For human-made sources, the recommended annual dose limits for individual members of the public are 1 mSv (100 mrem) for continuous exposures and 5 mSv (500 mrem) for infrequent exposures. Exposures of the public should be justified (no exposures without expectation of benefit) and should be "as low as reasonably achievable" (ALARA). It is also recommended that if a single site could potentially expose members of the public to more than 25% of the dose limit, the operator should ensure that the annual dose from all human-made sources does not exceed 1 mSv (100 mrem). This guidance is similar to the federal guidance proposed by EPA, which is discussed in chapter 7, but it does not apply to natural sources and, by extension, apparently not to TENORM.

NCRP recognizes that there are circumstances when natural background itself, or more especially "natural radiation sources enhanced locally by man's operations for selected purposes, can give rise (sometimes quite inadvertently) to annual exposures above the level of 1 mSv. It then becomes necessary to consider at what exposure level remedial action, which may only be possible at substantial societal cost, should be undertaken. Remedial action levels involve a balance of risk with many other socio-economic factors."

Once a remedial action level is established, action is expected when a level above it is found. NCRP recommends that, once a decision is made to take remedial action, the action not be limited to reducing the dose to just below the action level. Reduction to levels substantially below the remedial action level, following the ALARA principle, might be obtainable and appropriate.

Most of the discussion on remedial action levels for natural sources in NCRP Report No. 116 (NCRP 1993a) is focused on indoor radon for which an action level of 7×10^{-3} Jh/m^3 (2 WLM/y) is recommended. This action level is recognized as 10 times the national average for indoor radon and appears to be justified partly on the basis of feasibility, rather than attainment of an a priori numerical risk limit.[5] NCRP states that "a remedial action level must, therefore, be chosen for which the greatest risks are avoided but the societal impacts are not excessive" and NCRP goes on to state that "Therefore, NCRP has proposed a remedial action level which is based on excess lifetime risk being no more than ten times the excess lifetime risk associated with the average annual background level found in homes that is 7.0×10^{-3} Jh/m^3/y."

---

[5]Although it seems clear in NCRP Report No. 116 (NCRP 1993a) that the remedial action level for indoor radon is justified as a multiple of 10 of the average indoor radon level, and that societal effects are considered, a predecessor report (NCRP 1984c) justifies the same limit on the basis of numerical risk such that "an excess risk of death from lung cancer of 2% or more over a lifetime for the individual exposed to elevated or enhanced levels of radon daughters should be avoided."

For other natural sources, NCRP states "that in the case of other exposure from natural radiation sources, considerations similar to those applied to radon would appear to be reasonable. Since the average exposure to individuals in the United States from natural radiation sources, excluding radon, is approximately 1 mSv annually, it is recommended that remedial action be undertaken when continuous exposures from natural sources, excluding radon, are expected to exceed five times the average or 5 mSv (500 mrem) annually." As in the case of radon, the 5-mSv action level appears to be justified, at least partly, on the basis of feasibility rather than attainment of an a priori numerical risk. The committee has assumed that this remedial action level is applicable only to pre-existing situations for which remedial action is the only remedy, and not to future practices.

It is important to note that the 5-mSv remedial action level is a total dose (except of radon) from natural radiation, including naturally distributed sources and any increases in natural radiation dose caused by human activities (that is, TENORM). For example, in an area with a somewhat elevated background of 1.5 mSv (excluding radon) from naturally distributed sources, the recommended remedial action level would amount to 3.5 mSv for the TENORM contribution.

INTERNATIONAL COMMISSION ON RADIOLOGICAL PROTECTION

International Commission on Radiological Protection Publication 60 (ICRP 1991) contains the most recent ICRP recommendations on limitation of exposure to ionizing radiation. As with NCRP Report No. 116 (NCRP 1993a), recommendations are made for both workers and the public. Recommended limits for the public are given for "practices which involve the intended release of radioactivity to the environment from installations," but are explicitly avoided for cases where intervention is the only possible remedy. ICRP uses essentially the same principles of dose limitation as NCRP. Both require that practices leading to increased dose be justified as being beneficial. NCRP uses ALARA as a guiding principle, whereas ICRP uses the term "optimization" to mean essentially the same thing. ICRP positions on limitation of dose to the public are summarized below.

For "practices" that involve the intended release of radioactivity to the environment from installations, ICRP recommends an annual dose limit for individuals of 1 mSv (100 mrem). Radon in air and natural and artificial radioactivity already present in the environment are specifically excluded from this limit, which is the same as that recommended by NCRP.

In some situations, the sources, pathways, and exposed people are already in place when decisions about control measures are being considered. Any new control procedures will commonly constitute intervention. ICRP states that "An important group of such situations is that involving exposure to natural radiation," so its recommendations concerning intervention appear to be applicable to existing TENORM situations. However, the recommendations concerning intervention are not applicable to decommissioning of licensed operations involving naturally occurring radionuclides, such as the milling of uranium ores (Clarke 1995).

According to ICRP, intervention should be justified in the sense that it does more good than harm; net benefit should be optimized. The dose limits for the public recommended by ICRP are only for use in the control of practices. The use of ICRP dose limits or other predetermined dose limits as the basis for deciding on intervention might involve measures that would be out of proportion to the benefit obtained and would thus conflict with the principle of justification. ICRP recommends against the application of dose limits for deciding on the need for, or scope of, intervention unless doses approach those at which serious deterministic effects are caused.

For the case of radioactive residues from previous events, ICRP states that "the need for and extent of remedial action has to be judged by comparing the benefit of the reductions in dose with the detriment of the remedial work, including that due to doses incurred in the remedial work."

HEALTH PHYSICS SOCIETY

The Health Physics Society (HPS), a US professional organization of radiation-protection specialists, has issued policy statements on radiation dose limits for the general public (HPS 1992) and radiation standards for site cleanup and restoration (HPS 1993). Several of the recommendations are directly applicable to TENORM; others are indirectly applicable.

HPS Recommendations on Radiation Dose Limits for the General Public

HPS endorses the dose limits for human-made sources recommended by NCRP (1993a) and adopted by the Nuclear Regulatory Commission (1991), noting that they are sufficiently conservative for public-health protection, compliance can be verified by actual measurement, they can be achieved in most cases without sacrificing important public benefits, and they can be applied without discrimination to essentially all human-made sources.

HPS Recommendations on Radiation Standards for Site Cleanup and Restoration

Regarding radiation standards for site cleanup and restoration, HPS recommends that radiation-protection standards be based on health risks and be clearly related to quantities that can be measured, such as radiation exposure rates or radionuclide concentrations in soil, on equipment, or in buildings. Standards should be consistent with the fundamental principles recommended for all radiation-protection activities, including for example, that radiation doses be ALARA.

The seven specific recommendations made by HPS are summarized below.

- Remedial action should do more good than harm.

- For decisions on decommissioning strategies, the ALARA principle should be applied to the total radiation dose to society, including workers at the site as well as the general public.

- For unrestricted use of a restored site, HPS endorses the limit of 1 mSv for the total effective dose equivalent[6] (TEDE) to any member of the public in any one year from all nonmedical, human-made sources combined recommended by ICRP (1991) and NCRP (1993a). For site cleanup and restoration standards, HPS recommends that the dose limit be applied to all site-specific, nonoccupational sources, except indoor radon, including natural radionuclides whose concentrations have been enhanced by human activities.

- HPS recommends that a compliance screening level of 0.25 mSv (25 mrem) be applied to the mean annual TEDE to the "critical population group," defined as the most highly exposed homogeneous group affected by the restored site. If the mean

---

[6]Total effective dose equivalent is used by some organizations to emphasize that the sum of the contributions from external and internal sources is meant. This term is not a part of the recommendations of ICRP or NCRP. Effective dose equivalent, without the modifier total, is sufficient to imply contributions from external and internal sources.

annual TEDE to the critical group is likely to exceed 25 mrem, an evaluation should be made to ensure that no individual is likely to receive an annual TEDE exceeding 1 mSv (100 mrem) from all site-specific, nonoccupational sources, excluding indoor radon.

- HPS recommends that standards for site cleanup and restoration include an assessment screening level, below which further dose assessment is not required. For all site-specific, nonoccupational sources of radiation exposure, excluding indoor radon, HPS recommends an assessment screening level of 0.05 mSv (5 mrem) for an annual mean TEDE to the critical group.

- For unrestricted release of sites containing radium-226, radium-228, or thorium-232, HPS recommends a soil concentration limit of 0.2 Bq/g (5 pCi/g) above the normal concentration for the region to prevent excessive radon-222 or radon-220 concentrations in indoor air. It further recommends that the concentration be averaged over an area of 25-100 m^2 and a soil depth of 1 m. A screening level for the same nuclides of 0.04 Bq/g (1 pCi/g) above the normal concentration for the region, averaged over the same area and depth, is also recommended.

- HPS recommends that standards for site cleanup and restoration be based on probabilistic risk assessments designed to provide the best estimates of the distributions and uncertainties of doses that are likely to be received after restoration through the use of state-of-the-art, peer-reviewed, thoroughly documented mathematical models and computer codes.

NUCLEAR REGULATORY COMMISSION

The Nuclear Regulatory Commission is not authorized to regulate TENORM as defined in this study, because such materials exclude source, special nuclear, and byproduct materials as defined in the Atomic Energy Act and the Nuclear Regulatory Commission's licensing authority is derived entirely from the act (see chapter 6). However, the Nuclear Regulatory Commission has issued regulations and guidance for licensed radioactive materials that are potentially relevant to regulation of TENORM.

Regulations for Decommissioning of Licensed Facilities

In 1997, the Nuclear Regulatory Commission established radiologic criteria in 10 CFR Part 20 for decontamination and decommissioning of licensed nuclear facilities (62 FR 39058). The standards apply to all licensed facilities except uranium- and thorium-recovery facilities subject to 10 CFR Part 40, Appendix A (for example, mill tailing sites) and uranium-solution extraction facilities. The standards define conditions under which sites may be released for unrestricted use by the public or released under restricted conditions.

The standards specify that sites will be considered acceptable for unrestricted use if the annual effective dose equivalent to average individuals in the critical group does not exceed 0.25 mSv (25 mrem), including the dose from groundwater sources of drinking water, and the residual radioactivity has been reduced to levels that are ALARA. If the dose constraint of 0.25 mSv per year for unrestricted use is not reasonably achievable or if compliance with the dose constraint would result in net harm, sites will be considered acceptable for license termination under restricted conditions if provisions for legally enforceable institutional controls would provide reasonable assurance that the annual effective dose equivalent will not exceed 0.25 mSv (25 mrem) or the annual effective dose equivalent is ALARA and would not exceed 1 mSv (100 mrem) for unrestricted use if the institutional controls were no longer in effect or 5 mSv (500 mrem) if compliance with the 1-mSv criterion is not achievable, is prohibitively expensive, or would result in net harm. The kinds of institutional controls over contaminated sites that would be effective in limiting exposures of the public include government custodianship and land-use restrictions specified in deeds. The regulations also specify that a licensee must provide sufficient financial assurance to allow appropriate institutional controls to be maintained.

The standards described above apply to human-made radionuclides and to the levels of naturally occurring radionuclides above background, and the standards apply for 1,000 y after decommissioning.

Guidance on Disposal of Residual Thorium or Uranium

Earlier Nuclear Regulatory Commission guidance relevant to TENORM addresses disposal of residual thorium or uranium from past processing operations (Nuclear Regulatory Commission 1981). The guidance discusses several options for disposal of the residual radioactive materials that depend on the concentrations of uranium and thorium; materials containing higher concentrations require more-restrictive disposal methods and controls on land use at the disposal site. The particular option that has been used most often and is particularly relevant to regulation of TENORM is described below.

The Nuclear Regulatory Commission guidance specifies that residual thorium or uranium may be disposed of with no restrictions on burial method if the concentrations do not exceed:

- 0.4 Bq/g (10 pCi/g) for natural thorium or uranium with their decay products present and in equilibrium.
- 1.3 Bq/g (35 pCi/g) for depleted uranium.
- 1.1 Bq/g (30 pCi/g) for enriched uranium.

The Nuclear Regulatory Commission also intended that these concentration limits would define acceptable remediation of contaminated sites to permit unrestricted use by the public. The limits may still be used for remediation even though, as described in the previous section, the Nuclear Regulatory Commission has since issued more-general dose-based standards for unrestricted release of contaminated sites (Nuclear Regulatory Commission 1997a).

The concentration limits given above were based on two considerations (Nuclear Regulatory Commission 1981). First, the limit of 0.4 Bq/g (10 pCi/g) for natural uranium or thorium is simply an average of EPA's cleanup standards for radium in contaminated soil of 0.2 Bq/g (5 pCi/g) in the top 15 cm and 0.6 Bq/g (15 pCi/g) below 15 cm, as specified in Subpart B of the mill-tailings standards in 40 CFR Part 192 (see chapter 7). The cleanup standards for radium in contaminated soil at mill-tailings sites are relevant for natural uranium or thorium with their decay products present and in equilibrium because radium and its decay products are the most important radionuclides in these materials.

Second, the concentration limits for depleted and enriched uranium were intended to correspond to a limit on annual absorbed dose of alpha particles of 0.01 mGy (1 mrad) to the lungs or 0.03 mGy (3 mrad) to bone,[7] which EPA had used in developing proposed guidance on transuranium elements in the environment (EPA 1977). The concentration limits were derived from the assumed dose constraints on the basis of an analysis of inhalation and ingestion pathways for an assumed exposure scenario (Nuclear Regulatory Commission 1981).

As discussed later in this chapter, an exemption level or cleanup standard for ^{226}Ra of 1.1 Bq/g (30 pCi/g) is contained in some state regulations

---

[7]For purposes of radiation protection, the dose equivalent from irradiation by alpha particles often is assumed to be 20 times higher than the absorbed dose (NCRP 1993a; ICRP 1991).

for TENORM. This value also has been applied to unrestricted disposal of waste materials that contain uranium in sanitary landfills at DOE sites (for example, Lee and others 1996;1995; Kocher and O'Donnell 1987).

DEPARTMENT OF ENERGY

The Department of Energy (DOE) is responsible for regulating TENORM arising from any of its authorized activities (see chapter 6). DOE manages and disposes of wastes that contain TENORM in two ways, depending primarily on the volume of the waste material.

Management and Disposal of Small Volumes of TENORM

Current DOE requirements for management and disposal of radioactive waste (DOE 1988) specify that small volumes of waste containing TENORM may be managed as low-level waste. The waste volumes that may be so managed are not specified, but the volumes must be sufficiently small that the waste-acceptance criteria for the intended low-level waste disposal facility would be met. Thus, this option normally would be used for discrete sources.

In addition to the requirement for compliance with a limit on annual effective dose equivalent of 1 mSv (100 mrem) for individual members of the public from all DOE sources combined (DOE 1990), acceptable disposals of low-level waste at DOE sites are defined by the following performance objectives for the disposal facility (DOE 1988):

- A limit on annual effective dose equivalent for individual members of the public of 0.25 mSv (25 mrem) from all release and exposure pathways combined.

- Limits on releases of radionuclides to the atmosphere such that the requirements of EPA's 40 CFR Part 61 (see chapter 7) are met.

- Reasonable efforts to maintain releases of radioactivity to the general environment ALARA.

- A limit on effective dose equivalent for individuals who might inadvertently intrude onto the disposal site after loss of active institutional controls (assumed to occur 100 y after disposal) of 1 mSv per year (100 mrem per year) for continuous exposure or 5 mSv (500 mrem) for a single acute exposure.

- Protection of groundwater resources consistent with federal, state, and local requirements.

The performance objective for individual members of the public is consistent with a similar performance objective for commercial low-level waste disposal facilities that had been established by the Nuclear Regulatory Commission (1982a). The dose criteria in the performance objective for inadvertent intruders are the same as the dose limits from all DOE sources combined (DOE 1990) and are consistent with EPA's proposed federal guidance on radiation protection of the public (EPA 1994d) discussed in chapter 7. Finally, the performance objective for protection of groundwater resources often has been interpreted in terms of compliance with EPA's interim or proposed standards for radioactivity in drinking water (40 CFR Part 141) discussed in chapter 7.

Management and Disposal of Larger Volumes of TENORM

At DOE sites, larger volumes of waste containing TENORM (diffuse sources) that cannot be managed as low-level waste may be managed as uranium or thorium mill tailings (DOE 1988). Thus, these materials are intended for disposal at specially designated DOE sites or at mill tailings disposal sites established under the Uranium Mill Tailings Radiation Control Act.

The requirements for control of residual radioactive materials at DOE sites that apply to TENORM managed as mill tailings are contained in Order 5400.5 (DOE 1990). The requirements address release of contaminated property for unrestricted use by the public, interim storage of residual radioactive material, and long-term management of uranium, thorium, and their decay products.

In all situations involving management and disposal of residual radioactive material, the requirements for radiation protection of the public (DOE 1990) must be met. These include a limit on annual effective dose equivalent of 1 mSv (100 mrem) for individual members of the public from all routine DOE activities and exposure to the residual radioactive material. Higher doses from acute exposure are permitted if the annual effective dose equivalent averaged over a lifetime is not expected to exceed 1 mSv (100 mrem). In addition, residual radioactivity shall be reduced in accordance with the ALARA objective. As noted in the previous section, radiation protection requirements at DOE sites are consistent with EPA's proposed federal guidance on radiation protection of the public (EPA 1994d).

The requirements for release of contaminated property (land and structures) for unrestricted use by the public are the same as the guidelines for residual radioactivity developed in the Formerly Utilized Sites Remedial Action

Program (FUSRAP) and the Surplus Facilities Management Program (SFMP) (DOE 1997). Those guidelines include the following provisions, subject to the overriding requirement of compliance with the annual dose limit of 1 mSv (100 mrem) from all DOE sources combined:

- Limits on residual concentrations of radium and thorium in soil, airborne radon decay products in occupied or habitable structures on private property, and external gamma radiation level inside buildings or habitable structures on a site as given in Subpart B of EPA's mill tailings standards in 40 CFR Part 192.

- Limits on residual concentrations of other radionuclides in soil shall be derived from the annual dose limit of 1 mSv (100 mrem) in radiation-protection requirements using prescribed site-specific procedures and data.

- Limits on residual radioactivity on surfaces of structures and equipment on the basis of guidelines developed by the Nuclear Regulatory Commission (1982b; 1974).

- Limits on residual concentrations of radionuclides in air and water such that appropriate federal and state standards will be met.

The requirements for interim storage of residual radioactive materials also are obtained from the guidelines for FUSRAP and remote SFMP sites and include the following provisions:

- Limits on ^{222}Rn concentrations in air above background of 3.7 kBq/m^3 (100 pCi/L) at any point within a site, 1.1 kBq/m^3 (30 pCi/L) averaged over a year and over a site, and 0.11 kBq/m^3 (3 pCi/L) averaged over a year at any location outside a site.

- A limit on release rate of ^{222}Rn above background of 0.7 Bq/m^2 (20 pCi/m^2) per second on the basis of a similar provision in Subparts A and D of EPA's mill tailings standards in 40 CFR Part 192.

- Limits on radionuclide concentrations in groundwater or quantities of residual radioactive material as established in federal or state standards.

- Control and stabilization features designed to ensure, to the extent reasonably achievable, an effective life of 50 y, with a minimum life of 25 years.

Finally, the requirements for long-term management of uranium, thorium, and their decay products, which also are obtained from the guidelines for FUSRAP and remote SFMP sites, include the following provisions:

- Limits on the release rate of ^{222}Rn to the atmosphere and the concentration in air outside the boundary of the contaminated area as given in Subpart A of EPA's mill tailings standards in 40 CFR Part 192.

- Protection of groundwater in accordance with applicable DOE orders and federal and state standards.

- A design lifetime for control and stabilization features as given in Subpart A of EPA's mill tailings standards in 40 CFR Part 192.

- Control of access to the site and prevention of misuse of on-site residual radioactive material by appropriate administrative controls and physical barriers designed to be effective, to the extent reasonable, for at least 200 y.

In summary, DOE requirements for management and disposal of larger volumes of waste containing TENORM that cannot be managed as low-level waste are based primarily on two considerations:

- Compliance with the annual dose limit of 1 mSv (100 mrem) for individual members of the public from all routine DOE activities combined, including exposure to TENORM and other residual radioactive material, as specified in Order 5400.5 (DOE 1990).

- Compliance with provisions for residual radioactive material based primarily on EPA's standards for uranium and thorium mill tailings in 40 CFR Part 192.

In addition, airborne releases from DOE sites, including releases of TENORM, must comply with the provisions of EPA's 40 CFR Part 61.

DOE's requirements for radiation protection of the public and the environment are being revised (DOE 1993a). The requirements for management of property contaminated with residual radioactive material, including

TENORM, might be modified somewhat from the current requirements in Order 5400.5 (DOE 1990) discussed here.

STATE STANDARDS AND GUIDELINES

The Conference of Radiation Control Program Directors (CRCPD) was established in 1968 to serve as a common forum for state radiation-control programs in the United States to communicate with each other and with the many federal agencies that have radiation-protection responsibilities. The major technical work of CRCPD is accomplished through various working groups. In addition to about 50 smaller working groups, CRCPD has a special commission to develop suggested state regulations for the regulation and control of TENORM (CRCPD 1998). CRCPD has a long history of involvement in the TENORM issue. Its 1978 task-force report on natural-radioactivity contamination problems (CRCPD 1978), prepared in cooperation with EPA's Office of Radiation Programs, was one of the first assessments of the scope of the problem and of potential radiation-control measures.

Since 1990, CRCPD has published Suggested State Regulations for the Control of Radiation (SSRCR). SSRCR generally parallels federal radiation-protection regulations but contains additional provisions on subjects regulated at the state, but not federal, level. Such additional subjects include nonionizing radiation and x-ray use in medicine. CRCPD has drafted TENORM regulations as Part N of SSRCR (Reynolds 1995).

The February 1997 draft of Part N (CRCPD 1997) provides for the licensing of TENORM-generating industries and includes the following selected provisions (relevant sections of Part N are shown in parentheses):

- Operations, uses, or transfers of TENORM are to be conducted in a manner such that no member of the public will receive an annual total effective dose equivalent (excluding radon and its decay products) of 1 mSv (100 mrem) from all licensed sources, including TENORM sources (N.5.a). Part N does not specify what fraction of the 1-mSv (100-mrem) dose limit can come from the TENORM disposed of or released for unrestricted use; that determination is to be made by the implementing state in light of existing federal standards to protect the general public (N.5.e). However, generally applicable decontamination and decommissioning standards being developed by a CRCPD working group for Part D of SSRCR call for air and water releases and soil-contamination levels to meet a 0.15-mSv (15-mrem) annual dose constraint, in line with current EPA guidance (Luftig and Weinstock 1997).

- Use, transfer, or disposal of TENORM is to be conducted to prevent accumulation of radon in residential structures, schools, and other public buildings in concentrations exceeding 150 Bq/m^3 (4 pCi/L). Compliance with this standard may be demonstrated by imposition of institutional controls or adherence to building codes. Institutional controls may include deed restriction or notification, recorded in the property title or by the placement of permanent markers, that TENORM has been disposed of at the site (N.5.d).

- For radium-bearing TENORM, materials containing ^{226}Ra or ^{228}Ra at less than 0.2 Bq/g (5 pCi/g) are exempt from licensing (N.4.a.i).

- Purposeful dilution to render TENORM exempt is not allowed (N.4.a.iii).

- Land may not be released for unrestricted use where the soil concentration of ^{226}Ra or ^{228}Ra (averaged over any 100 m^2 and to a depth of 15 cm) exceeds 0.2 Bq/g (5 pCi/g). This limit and the exemption limit for licensing may be relaxed if dose assessment shows that the indoor-radon and other criteria in Section N.5 are met (N.7.b).

- The disposal methods used with uranium mill tailings regulated under 40 CFR 192 are generally acceptable for TENORM. However, other disposal methods meeting the basic criteria in Section N.5 might also be suitable, such as downhole disposal of some oilfield wastes, landfill disposal, and on-site disposal in conjunction with institutional controls. Cost-benefit analysis may be used to evaluate such alternative disposal options (N.8.a).

A wide variety of regulatory control is exercised over TENORM by states. Some states rely on general regulations for the control of radiation. Others have enacted or are considering regulations addressing TENORM site cleanup (specifically, radium concentration in soil; see table 9.1), contamination of equipment, and disposal options. The current state-by-state regulatory picture is given by Peter Gray & Associates (1997).

OTHER GUIDANCES FOR TENORM

Table 9.1. State regulations on TENORM; soil ^{226}Ra site cleanup standards (adapted from Peter Gray & Associates, 1997).

State	^{226}Ra cleanup standard Bq/kg (pCi/g)
Arkansas	200/15 (5/15)[a]
Colorado (proposed)	200/15
Georgia	200/15 with high radon factor[b]
	1,100/15 (30/15)[c] with low radon factor
Louisiana	200/15 or 1,100 if the effective dose equivalent to members of the public does not exceed 1 mSv (100 mrem) per year
Michigan (proposed)	200/15
Mississippi	200/15 with high radon factor
	1,100 with low radon factor
New Mexico	1,100/15
North Dakota	200
New Jersey	Variable—depending on concentrations and volumes. Annual dose less than 0.15 mSv (15 mrem)
Oklahoma (proposed)	1,100/15
Oregon	200/15
South Carolina	200/15 with high radon factor
	1,100/15 with low radon factor
Texas	200/15 with high radon factor
	1,100/15 with low radon factor
Conference of Radiation Control Program Directors	200/15

[a] "200/15" is 200 Bq/kg (5 pCi/g) of radium in soil, averaged over any 100 square meters and averaged over the 15 centimeters of soil below the surface.

[b] High radon factor is a radon emanation rate greater than 0.7 Bq (20 pCi) per square meter per second. This radon flux rate is the post-remedial action level specified for the disposal of uranium mill tailings in 40 CFR 192.02. Low radon factor is a radon emanation rate less than 0.7 Bq (20 pCi) per square meter per second.

[c] "1,100/15" is 1,100 Bq/kg (30 pCi/g) of radium in soil, averaged over any 100 square meters and averaged over the first 15 centimeters of soil below the surface.

INTERNATIONAL ATOMIC ENERGY AGENCY

The current International Atomic Energy Agency (IAEA) guidance on radiation protection of the public is contained in the basic safety standards (IAEA 1996a) discussed in chapter 6. Those standards include two provisions that are applicable to TENORM other than indoor radon.

First, paragraph 2.5 of the "Requirements for Practices" specifies that "exposure to natural sources shall normally be considered as a chronic exposure situation and, if necessary, shall be subject to the requirements for intervention, except that ... public exposure delivered by effluent discharges or the disposal of radioactive waste arising from a practice involving natural sources shall be subject to the requirements for practices ..., unless the exposure is excluded or the practice or the source is exempted." In that statement, practice and intervention have the same meanings as in ICRP recommendations discussed earlier in this chapter, requirements for practices include a limit on annual effective dose of 1 mSv (100 mrem) from all sources combined, and excluded exposures include exposures to most raw materials that contain unmodified concentrations of naturally occurring radionuclides. Thus, the standards encourage the regulation of some exposures to TENORM as practices and their being subjected to the annual dose limit of 1 mSv (100 mrem) from all sources combined. That approach differs from ICRP recommendations for interventions involving TENORM other than indoor radon discussed earlier, which do not include a dose limit. However, because ICRP had not developed a recommended dose limit for protection against exposure to natural sources, the IAEA's standards also include the statement that "the General Obligations for practices concerning protection against natural sources will be that exposure to natural sources, which is normally a chronic exposure situation, should be subject to intervention and ... the requirements for practices should be generally limited to radon, the exposure to other natural sources being expected to be dealt with by exclusion or exemption of the source or otherwise at the discretion of the Regulatory Authority." Thus, IAEA intends that exposures of the public to TENORM other than indoor radon be controlled in accordance with ICRP recommendations for interventions and that the dose limit for practices not be applied.

Second, IAEA's standards include recommendations for the exemption from regulatory control of TENORM containing low activity concentrations and low total activities, as indicated in the statements on the "Requirements for Practices" given above. The exemption levels for naturally occurring radionuclides were developed on the basis of the following considerations:

- The annual effective dose for individual members of the public from the exempted practice or source should not exceed about 10 µSv (1 mrem).

- The annual collective effective dose from the exempted practice or source should not exceed about 1 person-Sv (100 person-rem).

- A series of bounding exposure scenarios for use and disposal are assumed.

- The application of exemption to natural sources, where these are not excluded, is limited to the incorporation of naturally occurring radionuclides into consumer products or their use as a radioactive source or for their elemental properties.

The first two considerations are taken from previous IAEA recommendations on exemption principles (IAEA 1988). The last consideration indicates that the exemption levels for naturally occurring radionuclides apply only to intentional and beneficial uses of these radionuclides and therefore do not apply to residual radioactive materials or other wastes containing TENORM.

For the most important naturally occurring radionuclides other than radon, the recommended exemption levels are as follows:

- For natural uranium or thorium, an activity concentration of 1 Bq/g (27 pCi/g) and a total activity of 10^3 Bq (27 nCi).

- For ^{226}Ra, an activity concentration of 10 Bq/g (270 pCi/g) and a total activity of 10^4 Bq (270 nCi).

- For ^{228}Ra, an activity concentration of 10 Bq/g (270 pCi/g) and a total activity of 10^5 Bq (2.7 µCi).

- For ^{210}Pb, an activity concentration of 10 Bq/g (270 pCi/g) and a total activity of 10^4 Bq (270 nCi).

In each case, all decay products are assumed to be present and in equilibrium.

Especially for uranium, thorium, and radium, the recommended exemption levels given above are substantially higher than EPA's standards for cleanup of residual radioactive material at mill tailings sites, as given in Subpart B of 40 CFR Part 192 (see chapter 7). However, as noted previously, the recommended exemption levels for naturally occurring radionuclides do not apply to residual radioactive materials that contain TENORM.

COMMISSION OF THE EUROPEAN COMMUNITIES AND OTHER NATIONS

The Commission of the European Communities (CEC) has issued revised directives for radiation-protection standards that should be implemented by member states of the European Union by the year 2000 (see chapter 6). Title VII of the standards addresses substantial increases in exposure due to natural radiation sources. In particular, the standards call attention, first, to operations with and storage of materials not usually regarded as radioactive but that contain naturally occurring radionuclides and cause a substantial increase in exposures of the public, and second, to activities that lead to the production of residues not usually regarded as radioactive but that contain naturally occurring radionuclides and cause a substantial increase in exposures of the public. However, the CEC standards do not include any additional guidance on suitable approaches to regulating the identified exposure situations involving TENORM (such as guidance on a dose limit).

The committee has not endeavored to obtain information on standards for TENORM, especially in the form of a dose limit, that have been established in other countries. However, in discussions of ICRP recommendations for natural sources, Clarke (1996; 1995) noted that many countries have established an annual dose constraint for specific practices or sources of about 0.3 mSv (30 mrem) and that this dose constraint logically could be applied to unrestricted release of contaminated sites that were licensed for such activities as uranium mining or milling and then decommissioned. In response to this suggestion, the committee notes that an annual dose constraint of 0.3 mSv (30 mrem) corresponds to a concentration of ^{226}Ra in surface soil that is only about half the cleanup criterion for radium specified in Subpart B of EPA's standards for mill tailings in 40 CFR Part 192.

The European Union directives (Euratom 1996) differ from the IAEA basic safety standards (IAEA 1996a) by including in Title VII, a special provision that takes care of substantial increases in exposure due to natural radiation sources. Member states of CEC might reach different derived limits for ^{226}Rn that take into account peculiarities of industrial processes and specificity of industrial sites and waste repositories. By the year 2000, CEC member states will have to comply with the European Union directives (Euratom 1996).

TRANSFERABILITY OF STANDARDS

A ^{226}Ra concentration in soil of 0.2 Bq/g (5 pCi/g) appears frequently in federal and state regulations dealing with TENORM disposal or site cleanup. It derives from standard-setting during the early 1980s (40 CFR 192) associated with the Uranium Mill Tailings Radiation Control Act of 1978 (UMTRCA). UMTRCA governs remedial actions for uranium mill tailings disposal sites and discontinued uranium-milling facilities. It covers the disposal of the impounded tailings and other wastes associated with the extraction of uranium and thorium, and the dispersed tailings and associated wastes that have contaminated surrounding land, buildings, and soil (see chapter 7). The impounded tailings, classified as byproduct materials under the Atomic Energy Act, are typically of much higher activity concentration [about 300-1,000 pCi/g (11-37 Bq/g)] than the contaminated soils in the latter group and are intended to be maintained in perpetuity under federal or state control. The diffuse TENORM considered in this report typically has an activity concentration similar to that seen in the soils contaminated by dispersed tailings. The UMTRCA regulation addressing these soils (hereafter referred to as the soil-radium cleanup standard) calls for the concentration of ^{226}Ra in remediated sites not to exceed background levels by more than 0.2 Bq/g (5 pCi/g) in the uppermost 15 cm of soil or more than 0.6 Bq/g (15 pCi/g) in subsurface layers.

Alternative cleanup standards for contaminated sites covered by 40 CFR 192 that were considered, but not adopted, by EPA during the drafting of the regulations would have allowed an average ^{226}Ra concentration of 0.2, 0.6, or 1.1 Bq/g (5, 15, or 30 pCi/g) at all depths in the soil; 1.1 Bq/g (30 pCi/g) is the present cleanup standard in some state regulations. Under a conservative external-exposure model (one that assumes continuous exposure to an infinite quantity of material), a ^{226}Ra concentration in soil of 1.1 Bq/g (30 pCi/g) results in an estimated annual dose of 4.7 mSv (470 mrem) to the maximally exposed individual—close to the federal radiation-protection guidance for exposure of the general public from all sources combined that was then set at 5 mSv (500 mrem) (EPA 1982).

The Nuclear Regulatory Commission has applied the soil-radium cleanup standard to the decommissioning of active, Nuclear Regulatory Commission-licensed uranium mills (10 CFR 40; Appendix A). Additional examples of the regulatory extension of the soil-radium cleanup standard to waste materials other than uranium mill tailings can be seen in other remedial programs managed by DOE and EPA. DOE guidelines for soil cleanup at FUSRAP sites use this soil-radium standard (DOE 1997). EPA is using the standard from 40 CFR 192 as an applicable or relevant and appropriate requirement in establishing cleanup levels for CERCLA (Comprehensive Environmental Response, Compensation, and Liability Act) sites with ^{226}Ra contamination, as opposed to the annual dose constraint of 0.15 mSv (15 mrem) in current cleanup guidance (Luftig and Weinstock 1997). Finally, and of considerable practical importance to regulation of TENORM, the standard and variations on it discussed below form the basis of many existing and proposed state regulations dealing with TENORM.

The soil-radium cleanup standard has been extended broadly to encompass soil contamination by other uranium- and thorium-series radionuclides. For example, the dose associated with the soil-radium cleanup standard is being considered by the Nuclear Regulatory Commission as a benchmark for the cleanup of other radionuclides in connection with license termination at uranium and thorium mills and in situ leach (specifically, uranium solution extraction) facilities (Nuclear Regulatory Commission 1997c). A 1993 memo from the director of EPA's Office of Radiation and Indoor Air regarding the application of the 40 CFR 192 soil-radium cleanup standard to soils at a New Jersey FUSRAP site (and other non-UMTRCA sites) notes that the concentration criterion for surface soil was developed for the UMTRCA sites and that one would have to determine whether the risk scenarios at this FUSRAP site were sufficiently similar to those at UMTRCA sites to warrant the use of this health-based standard (EPA 1993a). This concern also applies to the non-fuel-cycle TENORM considered in this study.

The scientific validity of applying the soil-radium cleanup standard to TENORM wastes outside the nuclear fuel cycle, including industrial materials that are not classified as wastes, depends on the degree of similarity that such materials exhibit to the UMTRCA materials for which the standard was derived. Yet many non-fuel-cycle TENORM have initial mineralogies and processing histories different from uranium mill tailings. This results in materials in which the radionuclides can have different radon-emanation coefficients, solubilities, and bioavailabilities from uranium mill tailings. The zircon-bearing sands used in the metal-casting industry are a case in point. The occurrence of zircons as detrital minerals in river and beach sands is due to this mineral's high resistance to mechanical and chemical weathering. Leaching of radionuclides from zircons is low (Faure 1986). In contrast, radium and other radionuclides in uranium mill tailings are quite leachable (Landa 1982; 1980). Whereas uranium-mill tailings tend to have radon-emanation coefficients of about 10-40%, the values for zircons tend to be less than 5%. Thus, because of substantially lower leaching and radon emanation exhibited by zircon sands, the environmental mobility of radionuclides from these materials is much lower than that of radionuclides from uranium mill tailings. As a result, the potential for internal radiation exposure from a zircon source (such as leaching of radionuclides into groundwater and emanation of radon to the atmosphere) is considerably lower than the potential for internal exposure from UMTRCA materials.

However, whereas the low-release properties of some TENORM might limit the radon decay-product dose and internal dose contributions from other radionuclides, the external exposure must also be considered (table 2.8). If we consider a hypothetical, unshielded individual spending 100% of his or her time residing on homogeneous soil of infinite thickness, a ^{226}Ra concentration of 0.2 Bq/g (5 pCi/g), whether in the form of uranium mill tailings or any other radium-containing TENORM, will contribute more than half the 1-mSv (100-mrem) annual dose limit for public exposures permitted under the proposed federal radiation guidance (EPA 1994d). Incremental increases in radium concentrations in the soil of 0.24-0.4 Bq/g (6-10 pCi/g) would contribute a larger proportion of the 1-mSv (100-mrem) annual dose limit and would exceed the limit at the upper end. Shielding by an uncontaminated earthen cover and a decreased time of exposure would reduce the external dose from radium in the soil.

Finally, considering the 0.15-mSv (15-mrem) annual dose constraint deemed protective by EPA (Luftig and Weinstock 1997) and focusing only on external exposure, we note that this dose is equivalent to an incremental increase in the ^{226}Ra concentration in soil of about 0.04 Bq/g (1 pCi/g), that is, an amount about equal to the average background concentration. In view of the spatial variability (both areal and within the soil profile) in natural background levels and in view of sampling and analytic uncertainties, it will likely be difficult to implement a 0.15-mSv (15-mrem) soil-cleanup standard for radium, particularly when the contamination is only marginally above the local background.

10
Comparison of Current Guidances for TENORM in the Environment

INTRODUCTION

Chapters 7-9 of this report discussed guidances for naturally occurring radionuclides and TENORM in the environment developed by regulatory authorities in the United States and other countries and by national and international advisory organizations, such as the National Council on Radiation Protection and Measurements (NCRP), the Health Physics Society (HPS), the International Commission on Radiological Protection (ICRP), and the International Atomic Energy Agency (IAEA). For consistency with the purpose of this study, guidances developed by the Environmental Protection Agency (EPA) were considered separately from guidances developed by other organizations.

This chapter presents summary comparisons of EPA guidances specific to TENORM with the guidances for TENORM developed by other regulatory or advisory organizations. As discussed in chapter 1, TENORM refers to naturally occurring radioactive materials (NORM) not regulated under the Atomic Energy Act whose radionuclide concentrations or potential for exposures of individuals or populations have been increased by human activities.

In comparing EPA guidances for TENORM with guidances developed by other organizations, indoor radon is considered separately from other TENORM. This distinction is based on historical precedents in developing guidance on radiation protection. As described in chapter 1, it has been maintained essentially because the relationship between the increased risk of lung cancer and exposure to short-lived radon decay products in air can be estimated, with some uncertainty, directly from epidemiological studies in various groups of miners (ICRP 1993b; National Research Council 1988) without the need to estimate the dose to radiosensitive tissues of the lung from irradiation by alpha particles following inhalation intakes and the risk per unit dose from alpha particles. The availability of epidemiological data which directly links cancer risk with concentrations of short-lived radon decay products in air bypasses the need to consider the complexities and attendant uncertainties in describing physical and biological processes in the lung following inhalation of radionuclides, which are particularly important for alpha emitters (ICRP 1994). A dosimetric approach to estimating lung cancer risks from radon also requires assumptions—for example, the deposition of radon decay products in the respiratory tract and the particular target cells at risk—that may be difficult to verify. Thus, both the BEIR IV Committee (National Research Council 1988) and UNSCEAR (1993) did not endorse the use of dosimetric models for calculating risks of lung cancer from exposure to radon. Radon is unique because for no other radionuclides, including other naturally occurring alpha emitters (radium, uranium, and thorium), can a complete characterization of cancer risk be obtained without estimating the dose per unit exposure and the risk per unit dose.

INDOOR RADON

Guidances for mitigation of exposures to indoor radon developed by EPA, regulatory authorities in other countries, and national and international advisory organizations are discussed in chapter 8. This section presents a summary comparison of EPA guidances for indoor radon with those developed by other organizations.

Guidances for mitigation of radon in homes are summarized in table 10.1. This summary indicates that EPA's recommended mitigation level lies toward the lower end of the range of values encompassed by the guidances developed by other regulatory and advisory organizations. Some of the reasons for the differences are as follows.

All regulatory or advisory organizations that have developed guidance for radon in homes have assumed about the same risk per unit exposure to short-lived radon decay products. As summarized in chapter 8, on the basis of the assumption of a relative-risk model, there are some differences in the risk per unit exposure assumed by various organizations, arising from such factors as differences in the time at which the data were evaluated (that is, differences in the observed number of lung cancers in study populations) and differences in the models for projecting future risks in study populations for people who are still alive. However, those differences are not large, and the differences among the various guidances do not reflect substantial differences of scientific opinion about risks posed by exposure to indoor radon. Rather, the differences among

Table 10.1. Comparison of EPA guidance for mitigation of radon in homes with guidances developed by other organizations[a]

Organization	Radon concentration (Bq/m^3)	Comments
EPA and DHHS (1994)	150	Mitigation at 70-150 Bq/m^3 is recommended if concentration can be reduced to below 70 Bq/m^3
Other nations	20-1,000	Range of values encompassed by guidances developed by many nations
NCRP (1993a; 1984c)	370	Recommended mitigation level is exposure to short-lived radon decay products at 7 x 10^{-3} Jh/m^3/y (2 WLM/y); radon concentration assumes that decay products are in 40% equilibrium with radon
ICRP (1993b)	200-600	Mitigation is justified for levels above about 600 Bq/m^3; range of values represents optimized levels based on application of ALARA objective expected to be appropriate in most countries
Other advisory organizations	100-600	Range of recommendations by Commission of the European Communities (CEC), International Atomic Energy Agency (IAEA), and Nordic Radiation Protection Institute

[a] Adapted from information presented in table 8.4.

the guidances result primarily from such factors as differences in average radon levels in homes in various countries, in judgments about maximum tolerable risks posed by exposure to indoor radon or levels of indoor radon that are reasonably achievable with available technologies for mitigating exposures, and in the population groups of concern in establishing the guidances.

The differences between EPA guidance for radon in homes and NCRP and ICRP recommendations are of particular interest. As noted above, these differences do not result from substantial differences in assumptions about the risk posed by exposure to indoor radon. NCRP and ICRP recommendations were based primarily on judgments about the maximum tolerable risk or dose to individuals, and the intent was mainly to mitigate exposures of the relatively few individuals who experience the highest risks (ICRP 1993b; NCRP 1984c). NCRP's recommended mitigation level was based on an assumption that lifetime risks of fatal lung cancers greater than 0.02 from exposure to radon in homes should be avoided. NCRP also recommended that levels of indoor radon be reduced below the mitigation level in accordance with the ALARA objective (as low as reasonably achievable). Therefore, actions to reduce levels of indoor radon, once undertaken, should result in radon exposures substantially below the recommended mitigation level. ICRP's recommended mitigation level was based on the assumptions that the annual effective dose from exposure to indoor radon should not exceed about 10 mSv, which corresponds to a lifetime risk of fatal cancers of about 0.04, and that the optimized annual effective dose, based on application of the ALARA objective, should be in the range of about 3-10 mSv, taking into account the various situations in different countries.

The EPA guidance also is concerned with protection of individuals who experience the highest exposures to indoor radon. However, the EPA guidance was based for the most part on considerations of average levels of radon in homes and a cost-benefit analysis of reducing these levels with available technologies (EPA and DHHS 1994); that is, the guidance reflected a greater emphasis on reducing exposures to radon in the greatest number of homes. Thus, the difference between EPA's mitigation level and the values recommended by NCRP and ICRP is explained in part by a difference in emphasis—reducing risks to individuals versus reducing risks to whole populations.

Many regulatory or advisory organizations also have developed separate guidance on mitigation of indoor radon in above-ground workplaces (excluding workplaces involving operations of the nuclear fuel cycle) and schools. These guidances are summarized in table 10.2. With the notable exception of EPA guidance for radon in schools, the guidances for workplaces or schools usually are substantially higher than the corresponding guidance for

Table 10.2. Comparison of EPA guidance for mitigation of radon in above-ground workplaces and schools with guidances developed by other organizations[a]

Organization	Radon concentration in workplaces (Bq/m^3)	Radon concentration in schools (Bq/m^3)	Comments
EPA and DHHS (1994)		150	EPA has not developed separate guidance for workplaces
Other nations	20-3,000	20-1,000	See comment in table 10.1
NCRP			NCRP has not developed separate guidances for workplaces and schools
ICRP (1993b)	500-1,500	500-1,500	See comment in table 10.1; increase in mitigation level by factor of 3 compared with homes reflects difference in residence times
IAEA (1996a)	1,000	1,000	Average of range of mitigation levels recommended by ICRP

[a] Adapted from information presented in table 8.5. Above-ground workplaces exclude those involving operations of nuclear fuel cycle.

homes. The increase is based on consideration of the lower residence times in workplaces and schools than in homes. Although EPA recognized the difference in residence times, the guidance for radon in homes was applied to schools primarily on the basis of a judgment that the mitigation level in homes is reasonably achievable in schools, taking into account the existing levels of radon in schools and a cost-benefit analysis of reducing these levels (EPA and DHHS 1994). It is also possible, although not yet demonstrated, that children, who are the primary population group of concern in schools, experience substantially higher radon risks than adults.

TENORM OTHER THAN INDOOR RADON

This section presents a summary comparison of EPA guidances for TENORM other than indoor radon with relevant guidances developed by other federal agencies (the Nuclear Regulatory Commission and the Department of Energy, DOE), state organizations, and advisory organizations (NCRP, HPS, and ICRP). As noted above, this comparison does not consider guidances for naturally occurring radionuclides that apply to operations of the nuclear fuel cycle or the management and disposal of uranium or thorium mill tailings, which are regulated by EPA, the Nuclear Regulatory Commission, and DOE under the Atomic Energy Act. EPA guidances for TENORM other than indoor radon are discussed in chapter 7, and the relevant guidances developed by other organizations are discussed in chapter 9.

For consistency with the presentations in chapter 7, guidances for TENORM other than indoor radon are divided into two categories: those which apply to multiple sources of exposure combined, including sources other than TENORM, and those which apply only to specific sources or practices involving TENORM.

Guidances on Radiation Protection of the Public Applicable to TENORM

Guidances for TENORM other than indoor radon that apply to multiple sources of exposure combined are summarized in table 10.3. These guidances generally have been developed in the context of radiation-protection standards for the public (see chapter 7). The following points should be noted in comparing them.

Table 10.3. Comparison of EPA guidance on radiation protection of the public applicable to TENORM other than indoor radon with similar guidances developed by other organizations[a]

Organization	Annual dose equivalent (mSv)	Comments
EPA (1994d)	1.0	Proposed revision of federal guidance; dose limit for all controlled sources combined, including TENORM and human-made radionuclides but excluding indoor radon and medical exposures
FRC (1960)	5.0[b]	Existing federal guidance; dose limit for all controlled sources combined, including TENORM and human-made radionuclides but excluding indoor radon and medical exposures
DOE (1990)	1.0	Dose limit for all routine DOE activities and exposure to TENORM under DOE control
CRCPD (1997)	1.0	Draft suggested state regulations; dose limit for all controlled sources combined, including TENORM and human-made radionuclides but excluding indoor radon and medical exposures
NCRP (1993a)	5.0	Recommended remedial-action level for natural sources including natural background, TENORM, and naturally occurring radionuclides from nuclear fuel cycle but excluding radon
ICRP (1991)		Control of exposures to TENORM normally considered to involve interventions; site-specific approach based on principle of justification and application of ALARA objective, but without use of predetermined dose limit, is recommended

[a] Adapted from information presented in chapters 7 and 9.
[b] Dose limit applies to maximally exposed individuals. Average individual dose in exposed populations should not exceed one-third of dose limit for maximally exposed individuals.

First, this committee has assumed that the annual dose limit of 1 mSv (100 mrem) in EPA's proposed federal guidance on radiation protection of the public (EPA 1994d), rather than the annual dose limit of 5 mSv (500 mrem) in the existing federal guidance (FRC 1960), represents EPA's current views on the maximum tolerable dose from exposure to all controlled sources combined, even though the proposed revision of the guidance has not been issued in final form.

Second, the annual dose limit of 1 mSv (100 mrem) for all controlled sources combined, including TENORM and human-made radionuclides but excluding indoor radon and medical exposures, in EPA's proposed federal guidance on radiation protection of the public (EPA 1994d) is the same as the dose limit for all human-made sources combined recommended by the ICRP (1991) and NCRP (1993a). However, the NCRP and ICRP recommendations do not apply to TENORM and so are not given in table 10.3.

Third, the NCRP recommendation in table 10.3 applies only to natural sources (including natural background radiation), TENORM as defined in this study, and presumably uranium and thorium mill tailings, but it does not apply to human-made radionuclides. Therefore, the Federal Radiation Council and EPA guidances in table 10.3 that apply to TENORM other than indoor radon and to human-made radionuclides but not to natural background are not directly comparable with NCRP's recommended remedial-action level for natural sources. However, given that the average annual dose from natural background excluding radon is about 1 mSv (100 mrem) (see table 2.10), NCRP's remedial action level of 5 mSv (500 mrem) normally would allow considerably higher exposures to TENORM than EPA's proposed primary dose limit of 1 mSv (100 mrem).

Finally, as emphasized in chapter 7, acceptable radiation protection of the public is not defined solely in terms of compliance with a dose limit from exposure to all controlled sources combined. Rather, a basic principle of radiation protection is that the ALARA objective should be applied in reducing doses below the limit, and the ALARA objective is an integral part of all guidances listed in table 10.3.

Guidances for Specific Sources or Practices Involving TENORM

Guidances that apply only to specific sources or practices involving TENORM other than indoor radon are summarized in table 10.4. As discussed in chapter 7, standards for specific sources or practices are an important means of ensuring compliance with radiation-protection standards for all controlled sources combined, including the primary dose limit and the ALARA objective. The guidances summarized in table 10.4 differ from those listed in table 10.3 in

Table 10.4. Comparison of EPA guidances for specific practices or sources involving TENORM other than indoor radon with similar guidances of other organizations[a]

Organization	Standard	Comments
EPA (Luftig and Weinstock 1997)	Annual dose equivalent of 0.15 mSv	EPA has not developed regulations specific to TENORM other than indoor radon that would apply to all sources of exposure, but some TENORM is regulated under CERCLA; standard is EPA guidance on complying with goal for lifetime cancer risk of 10^{-4} under CERCLA (40 CFR Part 300).
EPA (40 CFR Part 141)	Radium at 0.2 Bq/L and gross alpha-particle activity at 0.6 Bq/L in drinking water	Standards for drinking water at the tap, including background, developed under Safe Drinking Water Act are used as guidance for protection of groundwater resources (EPA 1991b).
EPA (40 CFR Part 440)	Dissolved ^{226}Ra at 0.4 Bq/L, total ^{226}Ra at 1.1 Bq/L, and uranium at 4 mg/L in liquid discharges	Standards for daily effluents in liquid discharges from mines or mills used to process specified ores; limits on average concentrations in daily effluents for 30 consecutive days are factor of 2 or 3 lower.
EPA (40 CFR Part 61)	Annual dose equivalent of 0.1 mSv	Standard for airborne emissions from specified sources developed under Clean Air Act and based on lifetime cancer risk of 10^{-4}; limits on emissions of ^{210}Po from some sources also are specified.
Nuclear Regulatory Commission (1981)	Uranium or thorium at 0.4-1.3 Bq/g in soil	Guidance on unrestricted disposal of waste does not apply to TENORM, but some state guidances for TENORM are consistent with Nuclear Regulatory Commission guidance.

[a] Adapted from information presented in chapters 7 and 9.

Table 10.4. (Continued)

Organization	Standard	Comments
DOE (1990)	Radium and thorium at 0.2–0.6 Bq/g in soil 0.03 WL for indoor-radon decay products Indoor gamma radiation level of 20 μR/h	For release of contaminated property, limits on concentrations of radium and thorium in soil above background, radon in structures including background, and indoor gamma-radiation level above background as in EPA standards for uranium or thorium mill tailings (40 CFR Part 192).
HPS (1993)	^{226}Ra, ^{232}Th, or ^{228}Ra at 0.2 Bq/g in soil	Concentration limit above background for unrestricted release of contaminated sites based on cleanup criteria for soil in EPA standards for uranium or thorium mill tailings (40 CFR Part 192).
States	^{226}Ra at 0.2–1.1 Bq/g	Range of cleanup standards and exemption levels for ^{226}Ra in several state regulations; limits are based on cleanup criteria for ^{226}Ra in soil in EPA standards for uranium or thorium mill tailings (40 CFR Part 192) and additional considerations of radon emanation rate from contaminated materials.
New Jersey	Annual dose equivalent of 0.15 mSv	Standard for cleanup and exemption based on EPA guidance for complying with goal for lifetime cancer risk of 10^{-4} under CERCLA.
CRCPD (1997)	^{226}Ra or ^{228}Ra at 0.2–0.6 Bq/g in soil	Draft suggested state regulations for unrestricted release of contaminated land, provided that annual dose limit of 1 mSv from all controlled sources combined is met; limits are based on cleanup criteria for radium in soil in EPA standards for uranium or thorium mill tailings (40 CFR Part 192).
CRCPD (1997)	Appropriate fraction of limit on annual dose equivalent of 1 mSv for all controlled sources combined	Draft suggested state regulations for disposal or unrestricted release of contaminated materials.

that they do not apply to human-made radionuclides or to natural sources other than TENORM.

Guidances specifically for TENORM have been developed only by EPA, DOE, and state organizations. However, this committee also considers that the Nuclear Regulatory Commission guidance on natural uranium and thorium in soil and the HPS recommendation on cleanup standards for radium and thorium in soil are relevant to TENORM. In current ICRP (1991) and NCRP (1993a) recommendations, exposures to specific sources or practices involving TENORM other than indoor radon are considered only in the context of guidances on radiation protection of the public, which were considered in the previous section and summarized in table 10.3. The following points should be noted in comparing the guidances in table 10.4.

First, as noted in chapters 6 and 7, EPA's current approach to regulating TENORM other than indoor radon is rather fragmentary because no standard or set of standards applies to all potentially important exposure situations. However, in addition to requirements for complying with any applicable regulations—including those developed under the Safe Drinking Water Act, Clean Water Act, and Clean Air Act (table 10.4)—this committee has assumed that an annual dose constraint of 0.15 mSv (15 mrem) (Luftig and Weinstock 1997) represents EPA's current views on acceptable exposures to TENORM from any source.

Second, it is difficult to compare EPA guidance (Luftig and Weinstock 1997) expressed in terms of dose with standards for radium-226 expressed in terms of activity per unit mass developed by other organizations (see table 10.4), because the dose from exposure to materials containing a given concentration of ^{226}Ra can depend on the volume of the source. For example, for large volumes of contaminated surface soil, the annual dose from external exposure corresponding to the state and Conference of Radiation Control Program Directors (CRCPD) standards for ^{226}Ra expressed in terms of activity concentrations would be about 0.6 mSv (60 mrem) or greater for continuous occupancy (see chapter 7). Therefore, for large-volume sources, the annual dose constraint of 0.15 mSv (15 mrem) in EPA guidance should be considerably more restrictive than the state or CRCPD standards, with the exception of the standard in New Jersey. However, for much smaller sources, the external dose corresponding to a given concentration of ^{226}Ra could be reduced considerably. For example, at the outer surface of a steel pipe that contains ^{226}Ra contamination at 0.6 Bq/g (15 pCi/g) on the inside wall, Bernhardt and others (1996) estimated that the exposure rate would be about 2 μR/h. If the contamination is assumed to be represented by a line source, for which the dose varies inversely with the distance from the source, and if the outer surface of the pipe is assumed to be about 1 cm from the source, the exposure rate at a nominal distance of 1 m would be 0.02 μR/h, and the annual dose equivalent for continuous exposure at this distance, taking into account that 1 R is about 7 mSv (0.7 rem) (ICRP 1987a), would only be about 1 µSv (0.1 mrem). Similar comparisons with EPA's dose constraint would apply to the Nuclear Regulatory Commission, DOE, and HPS guidances in the form of activity concentrations.

Third, in regard to EPA guidance for cleanup of Comprehensive Environmental Response, Compensation, and Liability Act (CERCLA) sites (Luftig and Weinstock 1997), this committee again emphasizes that the annual dose constraint of 0.15 mSv (15 mrem) from exposure to TENORM atcontaminated sites subject to remediation under CERCLA is a goal, rather than a regulatory limit, and the goal may be exceeded if compliance is not feasible. That consideration compounds the difficulties in comparing EPA guidance with the various state standards for cleanup of ^{226}Ra, which are interpreted as regulatory limits. The dose criterion in EPA guidance essentially may be regarded as an upper bound on de minimis (negligible) dose, rather than a limit that must be met for specific sources or practices (see chapter 7).

Fourth, some state regulations for ^{226}Ra include both cleanup standards and exemption levels, and the two usually are essentially the same. Thus, in effect, these regulations specify that acceptable exposures to TENORM other than indoor radon can be no higher than exposures that do not warrant regulatory control.

Finally, the proposed Part N of the suggested state regulations (CRCPD 1997) also specifies conditions, not shown in table 10.4, for unrestricted release of facilities and equipment contaminated with TENORM, including limits on surface contamination for alpha and beta or gamma activity and limits on external radiation due to surface contamination. As described in chapter 9, such contamination limits also are specified in DOE requirements for unrestricted release of facilities and equipment contaminated with TENORM (DOE 1990). However, those criteria, especially the limits on surface contamination, are not clearly related to dose and so are not easily compared with the dose constraint in EPA guidance.

Bases For Differences In Guidances

The information discussed in the previous two sections and summarized in tables 10.3 and 10.4 indicates that current EPA guidances for TENORM other than indoor radon often are substantially more restrictive than similar guidances developed by other organizations. That is especially the case if one compares EPA's dose limit for all sources combined with NCRP's remedial action level for natural sources (table 10.3) or EPA's preferred dose constraint for individual sources with other guidances in the form of activity concentrations of radionuclides (table 10.4). However, as in the case of indoor radon discussed previously, all regulatory or advisory organizations that have developed guidances for TENORM other than indoor radon assume essentially the same risk per unit dose from uniform exposure of the whole body on the basis of estimates by expert groups (ICRP 1991; National Research Council 1990).

As discussed in chapter 11, EPA has developed methods of risk assessment for exposure to radionuclides other than radon that differ from the approaches normally used by other organizations, and the differences in estimated risks are particularly important for internal exposure to some naturally occurring radionuclides. However, the differences between EPA guidances for TENORM other than indoor radon, especially the annual dose limit of 1 mSv (100 mrem) for all controlled sources combined and an annual dose constraint of 0.15 mSv (15 mrem) for specific practices or sources, and the guidances developed by other organizations do not result from differences in methods of risk assessment for radionuclides. That is, EPA's approach to risk assessment, as it differs from the approach normally used by other organizations, was not an important factor in developing the current EPA guidances.

Thus, the differences between EPA guidances for TENORM other than indoor radon and the guidances developed by other organizations are not based on the differences of opinion about risks posed by exposure to TENORM. Rather, the differences between the guidances result in part from differences in judgments about acceptable risks from exposure to TENORM and differences in judgments about levels of TENORM in the environment that are reasonably achievable (see chapter 5). In addition, the guidances for TENORM in the form of concentration limits for radium and thorium in contaminated soil and other materials often were based primarily on existing EPA standards in 40 CFR Part 192 for cleanup of contaminated soil at uranium or thorium mill tailings sites (see chapter 9).

SUMMARY

This chapter has presented summary comparisons of guidances for controlling exposures of the public to TENORM developed by EPA with similar guidances developed by other organizations, including the Nuclear Regulatory Commission, DOE, state organizations, other countries, NCRP, HPS, ICRP, and IAEA. Guidances for indoor radon have been considered separately from guidances for TENORM other than indoor radon.

EPA's current mitigation level for indoor radon is somewhat lower than the values developed in most other countries or recommended by NCRP and ICRP. However, the differences in the guidances do not result from substantial differences of scientific opinion about the risks posed by exposure to indoor radon. Rather, they result primarily from such factors as differences in average radon levels in homes, differences in judgments about maximum tolerable risks to individuals or levels of radon that are reasonably achievable after mitigation, and differences in whether a guidance focuses primarily on reduction of risks to individuals receiving the highest exposures or on reduction of risks in the whole population.

In many cases, the current EPA guidances for TENORM other than indoor radon also appear to be more restrictive than the relevant guidances developed by state organizations, other federal agencies, NCRP, and HPS. However, direct comparisons of the various guidances are difficult and potentially misleading because of differences in whether exposures to natural background are included, the difference in concept between a regulatory goal and a limit, and the use of dose criteria in some guidances and activity concentrations of radionuclides in others. The differences between guidances do not reflect the differences in approaches to risk assessment for radionuclides used by EPA and other organizations. Rather, the differences in the guidances for TENORM other than indoor radon result primarily from differences in judgments about acceptable risk, differences in judgments about risks that are reasonably achievable, and judgments about the transferability of standards from one exposure situation to another.

An additional consideration of importance in comparing the guidances for TENORM summarized in this chapter is that the specified quantitative criteria usually are not the most important factor in determining doses and risks that would be experienced in any exposure situation. Rather, as discussed in chapter 7, actual doses and risks usually are determined primarily by application of the ALARA objective, largely without regard for any limits or goals for exposure to TENORM that might be specified in guidances.

11
Issues in Developing Guidances for TENORM

The primary purpose of this study has been to examine and report on the scientific and technical bases of guidances developed by the Environmental Protection Agency (EPA) for protection of the public from exposure to technologically enhanced naturally occurring radioactive materials (TENORM). The particular issue of concern to this study is whether the differences between EPA guidances for TENORM and those developed by other organizations are based on scientific and technical information or on policy decisions related to risk management. If there are differences in the scientific and technical bases of the various guidances, the relative merit of the different scientific and technical assumptions should be evaluated.

This chapter presents several summary discussions related to the purpose of the study, including discussions on:

- The question of whether the differences between EPA guidances for TENORM and those developed by other organizations have a fundamental scientific and technical basis.

- Specific areas in which the technical approaches to risk assessment of radionuclides developed by EPA differ from the approaches normally used by other organizations and the question of whether the differences have been important in developing guidances for TENORM.

- Specific areas in which the differences between EPA guidances for TENORM and those developed by other organizations are based on differences in policies related to risk management, rather than scientific and technical issues.

After those discussions, the chapter considers various alternatives for expressing guidances for TENORM and their implications for risk assessment, particularly with regard to the distinction between the risk-assessment issues that would need to be addressed in developing guidances and the issues that would be addressed in demonstrating compliance.

SCIENTIFIC AND TECHNICAL BASES FOR GUIDANCES

As summarized in chapter 10, there clearly are differences in the guidances for TENORM developed by EPA and similar guidances developed by other organizations, both for indoor radon and for TENORM other than indoor radon. Where there are differences, EPA guidances tend to be more restrictive.

However, this committee finds that the differences between EPA guidances for TENORM and those developed by other organizations do not have a scientific and technical basis. That conclusion is based on the observations that all organizations that have developed guidance on indoor radon have assumed essentially the same risk related to exposure to radon and its short-lived decay products on the basis of data obtained from studies of underground miners, and that all organizations that have developed guidances for TENORM other than indoor radon have assumed essentially the same risk related to uniform irradiation of the whole body on the basis of data obtained primarily from studies of the Japanese atomic-bomb survivors. Thus, for purposes of health protection of the public, including establishing guidances for acceptable levels of indoor radon and acceptable levels of exposure to TENORM other than indoor radon, all organizations have assumed essentially the same risks related to radiation exposure.

The lack of a scientific and technical basis for the differences between EPA and other guidances for TENORM does not imply that there are no differences in the technical approaches used in assessing risks related to radiation exposure. Indeed, this committee has learned of several such differences, as discussed in the following section. But differences in the technical approaches to risk assessment of radionuclides have not been the cause of the differences in the various guidances for TENORM.

This committee also notes that the various guidances for TENORM were developed at different times and that the basic assumptions about radiation risks have changed over time. For example, when the existing federal guidance on radiation protection of the public specifying an annual dose limit for individuals of essentially 5 mSv was issued (FRC 1961; 1960), quantitative information on the risks at low levels of exposure had not yet been developed by such groups as the National Council on Radiation Protection and Measurements (NCRP), the International Commission on Radiological Protection (ICRP), and the National Research Council. The genetic risk posed by radiation exposure was believed to be a greater concern than the cancer risk, and the limits on maximum and average annual doses in the federal guidance were based on a largely unquantifiable expectation that exposures below the dose limits would not result in an observable increase in cancers or genetic effects in exposed populations. By the time the proposed revision of the federal guidance was issued (EPA 1994d), the genetic risk was reduced in importance, cancer risks had been estimated from the atomic-bomb survivor data, and the estimated risks were used in conjunction with an assumption about the maximum tolerable risk posed by radiation exposure as a justification for lowering the annual dose limit for individuals to 1 mSv (see chapter 7).

Thus, the difference between the federal guidance (FRC 1961; 1960) and its proposed revision (EPA 1994d) clearly has a scientific basis. However, the issue of concern to this study is the difference between current EPA guidances for TENORM and those developed by other organizations, and this committee has assumed that the proposed revision of the federal guidance represents EPA's current views on requirements for radiation protection of the public. Therefore, because all current EPA guidances for TENORM and the guidances of other organizations have been developed or updated within the last decade, the assumptions about radiation risks have been essentially the same in all cases.

The committee was also asked to consider whether there is relevant scientific information that has not been used in the development of contemporary risk analysis of NORM. A particular concern is that some of the important naturally occurring radionuclides are parents of long decay chains involving complex mixtures of radioisotopes of different chemical elements, and that exposure to such mixtures of radionuclides might necessitate novel approaches to methods of risk estimation.

The decay chains of some naturally occurring radionuclides—especially radium, uranium, and thorium—are considerably more complex than the decay chains of other radionuclides with regard to the number of decay products and chemical elements involved. However, contemporary methods of risk assessment that estimate doses and risks associated with ingestion or inhalation of radionuclides by allowing any decay products produced in the body to be redistributed and retained in the body according to the metabolic behavior characteristic of the particular chemical element take this added complexity into account by using the same methods that are applied to other radionuclides with many fewer decay products. Thus, there is no evident need for a different approach in dealing with the complex decay chains of some naturally occurring radionuclides.

More generally, the committee is not aware of any evidence that there should be differences in risks, and thus differences in approaches to risk assessment, associated with exposure to naturally occurring and human-made radionuclides. Indeed, if one accepts the view currently held by all regulatory and advisory organizations involved in radiation protection that estimates of absorbed dose in tissue are the fundamental physical quantities that determine radiation risks for any exposure situation (NCRP 1993a; ICRP 1991), then there is no plausible rationale for any differences in risks between naturally occurring and any other radionuclides, because absorbed dose in tissue depends only on the radiation type and its energy but not on the source of the radiation.

Thus, in general, there should be no difference between NORM and any other radioactive materials with regard to suitable approaches to estimating doses and risks related to external or internal exposure. However, because naturally occurring radionuclides are ubiquitous in the exposure environment, there might be an increased opportunity, compared with many human-made radionuclides, to use observational data on natural levels in different environmental compartments (such as soil, water, air, plants, and animals) and the fluxes between compartments to calibrate exposure pathway models for TENORM. In contrast, the ability to use such natural analogue data in exposure analysis must be tempered by the recognition that the physical and chemical forms of TENORM can be substantially different from those of the same elements in the natural environment, in which case observations on the behavior of radionuclides in natural systems might not be relevant to the exposure situation of concern.

DIFFERENCES IN TECHNICAL APPROACHES TO RISK ASSESSMENT

During this study, the committee examined a white paper on risk harmonization that had been prepared jointly by EPA and the Nuclear Regulatory Commission (Nuclear Regulatory Commission/EPA 1995). The white paper includes discussions on similarities and differences in the methods of risk assessment of radionuclides used by EPA and the Nuclear Regulatory Commission.

The primary purpose of this section is to discuss and comment on the differences between the EPA and Nuclear Regulatory Commission approaches to risk assessment of radionuclides, and to discuss the importance of these differences with regard to the development of guidances for TENORM. This section also discusses the issues of truncation of risk assessments in time and transferability of standards from one exposure situation to another, which are particularly important for TENORM other than indoor radon.

Differences Between Environmental Protection Agency and Nuclear Regulatory Commission Approaches to Risk Assessment

The Nuclear Regulatory Commission's approach to estimating risk posed by chronic radiation exposure of the public normally is based on ICRP recommendations on estimating doses per unit exposure and the risk per unit dose. The Nuclear Regulatory Commission estimates lifetime risks on the basis of estimates of annual doses that are the sum of the annual dose equivalent to the whole body from external exposure and the 50-y committed effective dose equivalent (ICRP 1977) from ingestion and inhalation of radionuclides. Lifetime risk is estimated by multiplying the annual effective dose equivalent from external and internal exposure by the assumed exposure time (for example, 70 y) and the nominal risk of fatal cancers caused by uniform whole-body irradiation of 5×10^{-2} per sievert (ICRP 1991). It is important to note that ICRP's nominal risk factor takes into account the age dependence of radiation risk in the whole population, which is based on data on the atomic-bomb survivors (ICRP 1991).

EPA has developed a methodologically more rigorous approach to assessing risk posed by chronic lifetime exposure to radionuclides, which is particularly important for internal exposure and differs in several respects from the simple approach described above.

First, EPA calculates the total risk by first calculating the risk in each organ irradiated, which is based on the calculated absorbed dose and an assumed risk per unit dose for that organ, and then summing the calculated risks for all organs (Puskin and Nelson 1995; EPA 1994c). Thus, in estimating risk, EPA does not use the calculated effective dose equivalent, with its assumption of nominal risks per unit dose equivalent for various organs (which are intended only to be approximate indicators of risk), multiplied by ICRP's nominal risk related to uniform whole-body irradiation. If there were no other differences, the two approaches would yield estimates of risk that differed only to the extent that the risks for different organs assumed by EPA were substantially different from the values assumed by ICRP in defining the effective dose equivalent, because EPA and ICRP assume nearly the same risk related to uniform whole-body irradiation. EPA's current risk estimate for whole-body irradiation (Eckerman and others 1998) is about 12% higher than ICRP's, but this difference is not significant.

EPA's approach described above gives different estimates of risk from the approach used by the Nuclear Regulatory Commission, which is based on the effective dose equivalent (ICRP 1977), and the current approach of ICRP (1991), which is based on the effective dose, because EPA estimates risks to specific organs that are not considered explicitly in calculating the effective dose equivalent (for example, organs of the gastrointestinal tract and the kidneys) and has assumed different risks for many important organs. EPA's risk estimate for bone is less than ICRP's current estimate (ICRP 1991) by a factor of about 5 primarily because EPA recognized an error in ICRP's estimate that is based on a confusion between dose to the radiosensitive endosteal tissues and average skeletal dose (Puskin and others 1992; Bair and Sinclair 1992). And EPA's estimated risk factors for some other organs (such as the stomach, which is not considered explicitly in the effective dose equivalent, and the lungs) differ substantially from ICRP's current estimates (ICRP 1991), primarily because the two organizations estimated risks in different populations with different organ-specific background risks as a function of age (EPA 1994c). EPA's risk estimates were developed for a US population, but ICRP's risk estimates were developed for an average of several national populations (ICRP 1991). In addition, in developing the tissue weighting factors for the effective dose equivalent (ICRP 1977) and the revised tissue weighting factors for the effective dose (ICRP 1991), ICRP used rounded and binned values of the risk for the different organs of concern—an approach that has not been used by EPA.

Second, in risk assessments of internal exposure to radionuclides with radioactive decay products, the Nuclear Regulatory Commission and EPA use different assumptions in calculating the dose due to ingrowth of decay products in the body after intake of the parent radionuclide. The Nuclear Regulatory Commission uses a model recommended previously by ICRP (1979) in which most decay products are assumed to be retained in the organs of deposition of the parent according to the retention function for the parent, even though the metabolic behavior of the decay products often is different from that of the parent. EPA has developed more-sophisticated models incorporating the physiologically based biokinetic models developed more recently by ICRP (1996; 1995; 1993a; 1989a), which assume that decay products are redistributed and retained in the body according to their own metabolic behavior.

Third, in estimating doses and risks related to exposure to alpha particles, the Nuclear Regulatory Commission uses a radiation quality factor of 20 to convert absorbed dose to dose equivalent for all irradiated organs, on the basis of the ICRP recommendation (ICRP 1991; 1977) of a single radiation quality factor for alpha particles that would apply to any tissue and stochastic biologic end point of concern. However, in estimating risks related to irradiation by alpha particles, EPA uses a relative biological effectiveness (RBE) of 1 for leukemia, 10 for breast cancer, and 20 for all other cancer sites, on the basis of organ-specific information on the risk per unit absorbed dose from alpha particles (EPA 1994c). Thus, EPA's risk estimates for irradiation of bone marrow by alpha particles are much less than the estimates used by the Nuclear Regulatory Commission, and the risk estimates for the breast are somewhat less.

Finally, as noted previously, the Nuclear Regulatory Commission normally estimates risks posed by chronic lifetime exposure on the basis of a calculated annual effective dose equivalent, which includes the 50-y committed effective dose equivalent from internal exposure for reference adults (ICRP 1977), an assumed time of exposure, and a nominal risk posed by uniform whole-body irradiation. For internal exposure, use of the effective dose equivalent in this way overestimates risk because it does not properly account for the dose received as a function of age at exposure and time after exposure, which are important concerns for chronic exposures of the public to long-lived radionuclides that are retained in the body for long periods. For chronic external exposure, EPA calculates lifetime risk essentially in the same way as the Nuclear Regulatory Commission because the dose is received at the time of exposure and, as noted previously, EPA assumes nearly the same nominal risk related to uniform whole-body irradiation.

However, for internal exposure, EPA estimates risks posed by chronic lifetime exposure of the public on the basis of age-specific dose rates and age-specific cancer risks rather than committed effective dose equivalents for adults and a nominal risk factor, as in the Nuclear Regulatory Commission approach (Eckerman and others 1998; Dunning and others 1984; Sullivan and others 1981; Dunning and others 1980). Particularly for internal exposure to long-lived radionuclides with long retention times in the body, EPA's approach more properly accounts for the dose received as a function of age at intake and time after intake. In essence, EPA estimates risk posed by chronic lifetime exposure as a convolution, over age at intake from birth to death and time after intake, of (1) the dose rate as a function of time after intake for any age at intake, as estimated with age-specific biokinetic and dosimetric models for ingestion and inhalation of radionuclides, (2) the risk at any future age per unit dose received at a given age, and (3) the probability of death from all competing causes as a function of age, as obtained from US life tables. The risk at any future age per unit dose received at a given age is estimated with an absolute-risk model for bone, skin, and thyroid but with a relative-risk model for all other organs (EPA 1994c). The relative-risk model incorporates age-specific background cancer risks from all causes in the US population.

Aspects of EPA's approach to risk assessment for radionuclides described above have been used in several regulatory activities, including risk assessments to support current standards for airborne emissions of radionuclides in 40 CFR Part 61 (EPA 1989d; 1989b), development of radionuclide-specific slope factors for use in risk assessments at contaminated sites subject to remediation under the Comprehensive Environmental Response, Compensation, and Liability Act (CERCLA) (EPA 1989c), and risk assessments to support the development of site-cleanup standards for radionuclides (Wolbarst and others 1996) (see chapter 7).

Comments on Differences Between Environmental Protection Agency and Nuclear Regulatory Commission Approaches to Risk Assessment

This committee offers the following comments on EPA's approach to risk assessment for chronic lifetime exposure of the public, especially internal exposure, as it differs from the approach normally used by the Nuclear Regulatory Commission for similar exposure situations.

First, EPA's approach should provide more realistic estimates of risk than the approach used by the Nuclear Regulatory Commission. All the factors described in the previous section—the use of organ-specific risks for many organs instead of risks based on the effective dose equivalent and a nominal risk from uniform whole-body irradiation, the use of updated biokinetic models in estimating dose from ingrowth of decay products in the body, the use of organ-specific RBEs for alpha particles, and the use of age-specific dose rates from internal exposure in conjunction with age-specific cancer risks—should result in more realistic estimates of risks associated with chronic lifetime exposure.

Second, the differences between EPA and Nuclear Regulatory Commission approaches to estimating radiation risks do not always result in substantial differences in estimated risks. When the dose is due primarily to external exposure or the internal dose is due primarily to short-lived radionuclides that are distributed nearly uniformly in the body and emit only low-LET radiations (photons and electrons), the differences in the risk estimates between using EPA and Nuclear Regulatory Commission approaches are insignificant, essentially because the risk posed by uniform whole-body irradiation recommended by ICRP (1991) takes into account the age dependence of both the radiogenic and background cancer risks. As noted previously, EPA's risk estimate for these cases (Eckerman and others 1998) is only slightly higher than ICRP's recommendation. The largest differences in estimated risks occur for internal exposure to long-lived, alpha-emitting radionuclides (such as thorium), which preferentially deposit in bone and have long retention times in the body. In those cases, the important tissues at risk are red marrow and bone, and the EPA approach can result in risk estimates for ingestion and inhalation exposure that differ from the risk estimates obtained with the Nuclear Regulatory Commission approach by more than an order of magnitude (Eckerman and others 1998), with EPA's risk estimates generally being lower.

Third, the Nuclear Regulatory Commission does not always estimate risks on the basis of the effective dose equivalent and a nominal risk related to uniform whole-body irradiation. It uses organ-specific and age-specific risk factors similar to EPA's assumptions in certain cases, including risk assessments of reactor accidents and other situations where the particular individuals at risk can be identified. Thus, the differences between EPA and Nuclear Regulatory

Commission approaches to risk assessment generally are important only for prospective and hypothetical chronic-exposure situations.

Fourth, EPA does not always use the more rigorous approach to risk assessment described in the previous section but, in some cases, uses the same approach as the Nuclear Regulatory Commission. EPA uses the more rigorous approach only in assessing risks for purposes of reaching decisions on rulemaking, such as decisions on the feasibility of establishing standards and the effects of alternative standards. However, when radiation standards are expressed in terms of dose equivalent, as is often the case, EPA uses the same approach to dose assessment as the Nuclear Regulatory Commission for purposes of demonstrating compliance. The dosimetric quantities currently used by EPA for compliance purposes are effective dose equivalents for reference adults (Eckerman and others 1988), which do not incorporate any of EPA's current assumptions for purposes of risk assessment involving organ-specific and age-specific doses and risks or biokinetic models for radionuclides and their decay products in the body. EPA has taken the customary approach incorporating ICRP recommendations in demonstrating compliance with standards expressed in terms of dose to maintain a stable and uniform regulatory framework for the nuclear community. Furthermore, in using an assumed limit on lifetime risk to derive a limit on annual effective dose equivalent from exposure to all radionuclides of concern for use in standards (see chapter 7), EPA uses essentially the same nominal risk per unit effective dose for any radionuclide as does ICRP (1991); but EPA does not take into account the results given by the more sophisticated models that continuous intakes of different radionuclides corresponding to a given annual committed effective dose equivalent for reference adults can correspond to substantially different lifetime risks.

Finally, given the differences between EPA and Nuclear Regulatory Commission approaches to risk assessment and the fact that EPA and the Nuclear Regulatory Commission use the same approaches in demonstrating compliance with radiation standards expressed in terms of dose, it is important to appreciate that the simplified approaches to risk assessment developed by ICRP (1991; 1977) and used by the Nuclear Regulatory Commission were believed to be reasonable for the needs of these organizations. ICRP and the Nuclear Regulatory Commission are concerned only with radiation protection, in which case dose provides a measure of risk; and the effective dose equivalent and, later, the effective dose were developed by ICRP to provide a reasonable surrogate for risk in any exposure situation. Furthermore, radiation protection is concerned with control of exposures without undue concern for the risks posed by actual exposure situations, provided that applicable dose limits and the ALARA (as low as reasonably achievable) objective are met. Therefore, for purposes of radiation protection, the use of effective dose equivalents and a nominal risk factor for uniform whole-body irradiation in estimating risks posed by chronic exposure to any radionuclides was believed to be satisfactory.

ICRP also recognized that there are radionuclide-specific differences in lifetime risks related to internal exposure for the same annual effective dose equivalent, but the simplified approach to estimating risk was judged to be satisfactory as long as these differences were within about a factor of 3 of the risk posed by external exposure. However, the recent EPA analyses indicating that more rigorous estimates of risk associated with chronic lifetime intakes can differ from estimates based on the effective dose equivalent and a nominal risk factor by substantially more than an order of magnitude for some radionuclides (Eckerman and others 1998) call into question the general suitability of using the effective dose equivalent (ICRP 1977) in estimating risk even for purposes of radiation protection.

Many of the differences between EPA and Nuclear Regulatory Commission approaches to risk assessment described in this section result from the use by the Nuclear Regulatory Commission, and other federal and state agencies, of the now outdated effective dose equivalent. ICRP has replaced this quantity with the effective dose (ICRP 1991), which incorporates a greater number of organs and updated information on organ-specific risks, and ICRP also has developed age-specific effective dose coefficients for inhalation and ingestion which incorporate the newer physiologically-based biokinetic models for radionuclides and their decay products (ICRP 1996; 1995; 1993a).

Thus, EPA's current approach to risk assessment differs from the approach to estimating risk based on current ICRP methods mainly in three respects. First, EPA estimates risk on the basis of age-specific absorbed dose rates and radiogenic risks, instead of committed effective doses and a nominal risk factor. Second, EPA estimates risk for a US population with a longer average lifespan and different background cancer risks as a function of age than ICRP, the risk factor for bone is the corrected value developed by EPA, and the cancer risk for breast is based on data for the United States, rather than the atomic-bomb survivors. Third, EPA uses different RBEs for alpha particles for leukemia and breast cancer than the standard radiation weighting factor of 20 used by ICRP.

The effect of the differences described above is that EPA's risk estimates are slightly higher than ICRP's for external exposure and for internal exposure to radionuclides with short retention times in the body, but EPA's risk estimates are substantially less than those obtained by using ICRP methods for internal exposure to some of the long-lived, alpha-emitting radionuclides occurring in TENORM. For ^{232}Th, for example, EPA's risk estimates for inhalation and ingestion are less than the estimates based on current ICRP methods by a factor of 4-5 (Eckerman and others 1998).

Importance of Approaches to Risk Assessment for Guidances for TENORM

The potential importance of the differences between EPA and Nuclear Regulatory Commission or ICRP approaches to risk assessment described above for the development of guidances for TENORM is difficult to evaluate. The concern here is only with guidances for TENORM other than indoor radon because all organizations use essentially the same assumptions in assessing risk related to indoor radon.

As summarized in tables 10.3 and 10.4, current EPA guidances for TENORM other than indoor radon (Luftig and Weinstock 1997; EPA 1994d) are expressed in terms of the annual effective dose equivalent. In these cases, EPA's more rigorous approach to risk assessment was not used in developing the particular dose criteria based on an assumed acceptable risk, but ICRP's nominal risk factor for all radionuclides (ICRP 1991) was used instead. Furthermore, the approach of calculating effective dose equivalents for reference adults (Eckerman and others 1988) would be used in demonstrating compliance with the guidance. Therefore, on the basis of the discussions in the previous two sections, the more rigorous approach to risk assessment would be used by EPA only for investigating the feasibility of any particular guidance for TENORM.

However, TENORM other than indoor radon has some unique characteristics among the various controlled sources of public exposure that could encourage a reexamination of the conventional approach to developing an annual dose criterion based on an assumed acceptable risk and ICRP's nominal risk factor. In contrast with human-made radionuclides from the nuclear fuel cycle, only a few radionuclides are of concern (isotopes of uranium, thorium, and radium and their shorter-lived decay products), and most of the radionuclides of concern are long-lived alpha-emitters that deposit in bone. Those are precisely the kinds of radionuclides for which the differences between EPA and Nuclear Regulatory Commission approaches to risk assessment are the most important and there is the greatest incentive to use a more rigorous approach to risk assessment to establish a dose criterion based on an assumed acceptable risk.

Furthermore, because only a few radionuclides are of concern in regulating TENORM, regulatory criteria conceivably could be expressed in terms of allowable concentrations of radionuclides in environmental media rather than dose. If the acceptable environmental levels are based on an assumed acceptable risk, they could be derived with EPA's more rigorous approach to risk assessment.

A factor that argues against this approach is that EPA's preliminary risk assessments for various scenarios of exposure to TENORM other than indoor radon indicate that external exposure often is considerably more important than internal exposure (EPA 1993b). Whenever that is the case, the use of more rigorous approaches to risk assessment for internal exposure would not have a substantial effect on the estimated risk because, as noted previously, EPA's approach to estimating risk related to external exposure is essentially the same as the approach used by the Nuclear Regulatory Commission and ICRP.

The extent to which rigorous approaches to estimating risk posed by chronic lifetime exposure are used in developing standards for radionuclides in the environment expressed in terms of dose or some other quantity, such as concentrations in environmental media, partly involves a judgment about the extent to which a standard should correspond to a particular risk related to exposure to any radionuclide of concern. Radiation protection of the public has not been unduly concerned with actual risks corresponding to a particular limit on annual dose as long as compliance with the dose limit and the ALARA objective is achieved. However, the emphasis on risk posed by radiation exposure clearly is increasing, owing in part to the increasing regulation of radiation exposures under environmental laws developed for hazardous chemicals, as well as radionuclides. For chemical carcinogens, regulations must be based on considerations of risk because there is no known surrogate for risk analogous to radiation dose. Therefore, in developing standards for radionuclides and chemical carcinogens, risk is the only available measure for comparing effects of exposure.

Given that the difficulties with conventional risk assessments for internal exposure based on ICRP recommendations are particularly important for long-lived alpha-emitting radionuclides, the development of guidances for TENORM other than indoor radon provides an opportunity to incorporate EPA's more rigorous approaches to risk assessment in all phases of standard development and demonstrations of compliance. The opportunity would need to be weighed against the desire to maintain a stable and uniform regulatory framework for controlling all radiation exposures of the public and the likelihood that there would not be a substantial effect on estimated risks when external exposure is considerably more important than internal exposure.

Other Issues in Risk Assessment and Guidance Development

This committee has considered two additional issues that are potentially important for risk assessment and the development of guidance for TENORM: truncation of risk assessments in time, and the transferability of standards from one exposure situation to another.

Truncation of Risk Assessments in Time EPA and the Nuclear Regulatory Commission generally truncate risk assessments in time for any situations involving management or disposal of materials that contain long-lived radionuclides, including management and disposal of mill tailings, disposal of high-level and transuranic wastes, and cleanup of contaminated sites and facilities (Nuclear Regulatory Commission/EPA 1995). As noted in chapter 7, risk assessments for these situations are performed only for 1,000 or 10,000 y. The issue of truncation of risk assessments in time is particularly important for long-lived radionuclides found in TENORM (radium, thorium, and uranium) because of their retention in the environment and the buildup of their radiologically important decay products.

Two justifications have been given for truncating risk assessments in time (Nuclear Regulatory Commission/EPA 1995). For some situations, available alternatives for managing future risks might not yield substantial differences in risks beyond some time. For example, the use of any type of engineered barrier in a disposal system presumably would be ineffective beyond some time. Therefore, there can be no basis for selection of the best alternative based on risk assessments beyond that time. For other situations, the alternatives might still differ in effectiveness in reducing risks in the future, but there is no scientifically reliable basis for distinguishing between the different capabilities. This justification was used, for example, in developing regulations for disposal of high-level and transuranic wastes: potential changes in the geologic environment were judged to render largely meaningless any predictions of effects more than 10,000 y hence (EPA 1985).

In this committee's view, the issue of whether it is reasonable to truncate risk assessments in time for establishing and implementing standards is not easily resolved, because either choice leads to conceptual difficulties. Truncation of risk assessments in time based on the justifications described above appears to violate the longstanding principle of radioactive-waste management that there should be no predictable future risks to human health that would be unacceptable today (IAEA 1995). That is especially true when the largest projected effects would not occur for tens of thousands of years or more, as is often the case for disposal of long-lived wastes at well-chosen sites. In such cases, truncation of risk assessments in time could give the appearance of arbitrarily ignoring the largest projected effects.

However, assessing risks essentially into eternity also seems unreasonable. Beyond some time, risk projections are likely to be largely meaningless in relation to actual effects on humans, given the inevitable changes in future living habits, changes in approaches to public health and improvements in medical care, and the likelihood of substantial changes in the geologic environment. Therefore, it might be quite unreasonable to base today's decisions about risk management on risks projected for the distant future and on present conditions. In addition, to the extent that decisions about risk management are based on cost-benefit analysis, an assumption that risks in the far future have the same value as risks in the near future appears to violate standard economic principles of discounting.

This committee believes that it is reasonable to truncate risk assessments in time for purposes of establishing standards and demonstrating compliance. However, we also believe that the selection of times for truncating risk assessments (such as 1,000 or 10,000 y) is largely a matter of judgment with a considerable degree of arbitrariness that should be acknowledged. Any estimates of times beyond which alternatives for managing risks would not result in substantial differences in risk or of times beyond which the geologic environment would be much different from present conditions clearly are highly uncertain.

However, the committee suggests that calculations of future risks should be carried out at least to the time of maximum projected effects, regardless of when they occur, even if the results are not used in establishing standards or in demonstrating compliance. Assessments of future risks over any time frames necessarily involve important assumptions that cannot be verified, and all projected risks for any times thus are somewhat arbitrary, but presentation of the full range of information about future risks should add value to risk assessment, even if not all the information is used in decision-making.

Transferability of Standards An important issue in developing guidances for TENORM is the question of whether it is appropriate to transfer standards developed for one exposure situation to other situations. This issue arises particularly because, as summarized in table 10.4 and discussed in chapter 9, some states have developed cleanup standards and exemption levels for radium-226 in the form of limits on concentrations in the range of 0.2-1.1 Bq/g (5-30 pCi/g) on the basis of cleanup standards for ^{226}Ra in contaminated soil at uranium mill tailings sites established by EPA in 40 CFR Part 192 (see chapter 7).

This committee generally supports the idea that standards for different exposure situations should be consistent to the extent reasonable, particularly standards expressed in terms of risk or dose. However, we also believe that considerable caution is warranted in transferring standards expressed in terms of activity concentrations of radionuclides from one exposure situation to another. The need for caution is exemplified by the standards for ^{226}Ra noted above.

Transfer of the cleanup standards for ^{226}Ra in contaminated soil at uranium mill tailings sites to other exposure situations involving ^{226}Ra might be inappropriate in several respects. First, the standards for mill tailings sites were based primarily on a judgment by EPA about levels of ^{226}Ra in soil that are reasonably achievable given the high background levels of ^{226}Ra in soil in the western United States, where uranium-ore deposits exist and the residual radioactive materials are found, and on the need to distinguish between naturally occurring ^{226}Ra in soil and ^{226}Ra arising from mill tailings by measurement of external radiation in the field. The standards for mill-tailings sites would not be appropriate for other exposure situations where the specified concentrations were not reasonably achievable.

Second, the cleanup standards for ^{226}Ra in contaminated soil at uranium mill tailings sites correspond to an annual dose that is an appreciable fraction of the annual dose limit of 1 mSv (100 mrem) for all controlled sources combined in EPA's proposed federal guidance on radiation protection of the public (EPA 1994d) (see chapter 7). Therefore, if a standard for ^{226}Ra that would apply to other exposure situations is intended to correspond to a limit on annual dose that is only a small fraction of the dose limit for all controlled sources combined, the standards for mill-tailings sites might not be appropriate, especially for large-volume sources.

Third, the external dose from localized sources of ^{226}Ra can be substantially less than the external dose from large-volume sources, such as a large extent of contaminated soil (for example, more than 100 m^2) with the same activity concentration. Therefore, using a single concentration standard for ^{226}Ra without regard for the size of the source could result in unduly restrictive regulation of localized sources if the standard is intended to correspond to a particular annual dose for any exposure situation.

Finally, as noted in chapter 7, the cleanup standards for ^{226}Ra in contaminated soil at uranium mill tailings sites are expected to correspond to concentrations of indoor-radon decay products of about 4×10^{-7} J/m^3 (0.02 WL). The assumed correspondence between radium concentrations in soil and levels of indoor-radon decay products applies only to materials in which the emanation rate of radon is similar to that in mill tailings. Therefore, if exposures to indoor radon are a potential concern, the radium standard for mill-tailings sites might not be appropriate for other situations where the emanation rate of radon from the materials in contaminated soil is substantially different from the emanation rate from mill tailings.

The issue of transferability of standards, especially standards in the form of concentration limits of radionuclides, is not easily resolved, primarily because radiation protection involves compliance with the ALARA objective, as well as a limit on dose or risk. Therefore, for example, the cleanup standards for ^{226}Ra in contaminated soil at uranium mill tailings sites could be applied to other exposure situations involving ^{226}Ra if the standards were reasonably achievable, even when there would be substantial differences in doses and risks. In transferring standards from one situation to another, it is important to investigate whether the standards are reasonably achievable for a variety of exposure situations of concern, especially if the doses and risks are substantially different. Differences in the physical and chemical forms of radionuclides in the different situations also need to be considered because the dose from internal exposure pathways can depend significantly on the form of the materials. Such considerations are important to ensure that standards that are reasonable for one exposure situation are not applied inappropriately to other situations.

POLICY-BASED DIFFERENCES IN GUIDANCES FOR TENORM

As indicated earlier in this chapter, this committee finds that the differences between EPA and other guidances for TENORM do not have a scientific and technical basis but, rather, result essentially from differences in policies for risk management. This section discusses a number of ways in which that is the case, including

- Selection of a limit on acceptable dose.
- Application of EPA's groundwater protection strategy to regulation of TENORM.
- Differences between the Nuclear Regulatory Commission's standards for decontamination and decommissioning of contaminated sties and EPA's preferred approach to radiation-site cleanup standards.
- EPA guidance on indoor radon vs. NCRP and ICRP recommendations.
- EPA guidance on dose limit for all sources of exposure combined vs. NCRP's recommendation on a remedial-action level for exposure to natural sources.
- The general treatment of natural background in establishing guidances.

All those considerations are potentially important in developing guidances for TENORM.

Limit on Acceptable Dose

The white paper on risk harmonization (Nuclear Regulatory Commission/EPA 1995) indicates that EPA and the Nuclear Regulatory Commission have fundamentally different views about a limit on acceptable risk related to radiation exposure and, therefore, about a limit on acceptable dose that might be included in guidances for TENORM other than indoor radon and for any other controlled sources of exposure. In particular, the white paper indicates that the annual dose limit of 1 mSv (100 mrem) specified in 10 CFR Part 20 (Nuclear Regulatory Commission 1991) is acceptable for individual Nuclear Regulatory Commission licensees, whereas the white paper and other guidance (Luftig and Weinstock 1997; EPA 1994d) indicate that in EPA's view, the dose from individual sources should normally be limited to substantially less than the annual dose limit from all sources combined of 1 mSv (100 mrem). EPA evidently favors an annual dose constraint for individual sources of 0.15 mSv (15 mrem), on the basis of the objective of achieving a lifetime risk of about 10^{-4} (Luftig and Weinstock 1997).

This committee offers the following comments on the issue of a limit on acceptable risk and, therefore, acceptable dose. First, the determination of an acceptable risk for any exposure situation clearly is entirely a matter of judgment (risk-management policy) which presumably reflects societal values. Inasmuch as EPA and the Nuclear Regulatory Commission have used essentially the same assumptions about the risks posed by radiation exposure in establishing radiation standards, it is clear that the determination of a limit on acceptable dose for any exposure situation also is entirely a matter of judgment. Therefore, any differences between the views of EPA and the Nuclear Regulatory Commission on an acceptable dose have no scientific or technical basis.

Second, a simple comparison of the Nuclear Regulatory Commission's annual dose limit for individual licensees of 1 mSv (100 mrem) (Nuclear Regulatory Commission 1991) with EPA's preferred annual dose constraint for individual sources of 0.15 mSv (15 mrem) (Luftig and Weinstock 1997) gives the impression that EPA's dose constraint would be considerably more protective of human health. However, this committee believes that such a comparison is quite misleading and, therefore, that the resulting impression is basically incorrect.

As emphasized in chapter 7, requirements for radiation protection of the public include implementation of the ALARA objective, as well as compliance with a dose limit for all controlled sources combined and dose constraints for individual practices or sources; and the ALARA objective is included in existing and proposed federal guidance on radiation protection of the public (see chapter 7) and the Nuclear Regulatory Commission's radiation-protection standards in 10 CFR Part 20. Thus, although the Nuclear Regulatory Commission allows annual doses as high as 1 mSv (100 mrem) for individual licensees, it also requires that all licensees implement an ALARA program. The effect of vigorous application of the ALARA objective has been that doses to the public achieved by nearly all licensees are only a few percent or less of the dose limit. Therefore, the practical effect of Nuclear Regulatory Commission requirements is that doses from nuclear facilities currently operating under Nuclear Regulatory Commission or Agreement State licenses are limited to levels that EPA would judge acceptable according to its preferred annual dose constraint of 0.15 mSv (15 mrem). The principal difference between EPA and Nuclear Regulatory Commission approaches to radiation protection is that EPA imposes dose constraints on particular classes of sources (such as operating nuclear fuel-cycle facilities) as a means of implementing the ALARA objective, whereas the Nuclear Regulatory Commission usually applies the ALARA objective only on a site-specific basis. That difference evidently has little practical importance in determining doses actually experienced.

Another important consideration in comparing EPA and Nuclear Regulatory Commission views on an acceptable dose is that EPA's preferred annual dose constraint for individual sources of 0.15 mSv (15 mrem) is a regulatory goal in the case of cleanups of radioactively contaminated sites, and the goal for cleanups can be waived if achieving the goal is not feasible (Luftig and Weinstock 1997). Therefore, as in the case of the Nuclear Regulatory Commission's annual dose limit of 1 mSv (100 mrem), EPA's dose constraint can be modified by ALARA considerations when applied to cleanup of contaminated sites. In this case, however, the important difference is that EPA's criterion can be relaxed, whereas the doses allowed by applying the ALARA objective to Nuclear Regulatory Commission licensees are always lower than the Nuclear Regulatory Commission's dose limit.

This committee also notes that it is somewhat misleading to label annual doses approaching 1 mSv (100 mrem) as "acceptable," even though they are allowed for individual Nuclear Regulatory Commission licensees under unusual circumstances. The ICRP (1991) has emphasized that annual doses approaching 1 mSv (100 mrem) are only "barely tolerable" and expects that doses usually can be reduced to well below barely tolerable levels by the use of source constraints at less than the dose limit and further site-specific applications of the ALARA objective. As noted previously, this is the case for most licensed sources. Doses are properly termed "acceptable" only when they are below the dose limit and are ALARA.

Application of the Environmental Protection Agency's Groundwater-Protection Strategy to TENORM

An important element of EPA's approach to protection of public health and the environment is its groundwater-protection strategy (EPA 1991b). The strategy defines protection of groundwater in terms of compliance with standards (maximum contaminant levels, MCLs) for radionuclides and other contaminants in public drinking-water supplies (see chapter 7), and it specifies that human activities today should not cause levels of contamination in groundwater that would entail later costs for removal if the groundwater is used as a source of drinking water. The application of MCLs in drinking water as standards for limiting contamination of groundwater from current operations, cleanup of contaminated sites, and waste disposal clearly has important implications for establishing guidances for any radioactive materials. That is especially so for TENORM because the radionuclides of concern occur naturally in all groundwaters.

As in the case of a limit on acceptable dose discussed in the previous section, application of EPA's groundwater-protection strategy in establishing guidances for TENORM and other radioactive materials clearly is a matter of risk-management policy. As discussed in chapter 7, MCLs for naturally occurring radionuclides in drinking water are based on considerations of existing levels in public drinking-water supplies and judgments about the cost effectiveness of reducing these levels with available technology for water treatment, but they are not based on an a priori judgment about an acceptable dose or risk related to exposure to radionuclides in drinking water. Furthermore, the judgments about levels of radioactivity that are reasonably achievable in public drinking-water supplies can change (EPA 1991a).

Given the basis for the MCLs for radionuclides in drinking water described above, it is clear that EPA's groundwater-protection strategy should be interpreted as defining a goal, rather than a requirement that must be met without regard for other circumstances. Therefore, application of the groundwater-protection strategy to guidances for TENORM is justified only to the extent that compliance with MCLs in groundwater that is a potential source of drinking water is reasonably achievable for the exposure situations of concern. In considering levels of contamination in groundwater that are reasonably achievable for any particular situation, it is important to consider not only the costs of achieving any particular levels in relation to projected health risks averted, but also such factors as the costs of primary treatment at the source in relation to potential future costs of secondary treatment by a water-supply system, the volume of groundwater that could be affected in excess of drinking-water standards, the period over which the projected effects could occur, and the ability of institutional controls to prevent future uses of contaminated groundwater and the associated costs of such controls.

Differences Between Nuclear Regulatory Commission and Environmental Protection Agency Approaches to Site-Cleanup Standards

The Nuclear Regulatory Commission recently issued standards for decontamination and decommissioning of licensed nuclear facilities (Nuclear Regulatory Commission 1997a) that define radiologic conditions for license termination and release of sites for unrestricted or restricted use by the public (see chapter 9). Sites generally are acceptable for unrestricted use if the annual effective dose equivalent from all exposure pathways, including use of groundwater as a source of drinking water, does not exceed 0.25 mSv (25 mrem) for a period of 1,000 y. Conditions for restricted release also are specified, and the standards allow for alternative criteria for license termination, provided that the annual effective dose equivalent from all sources combined does not exceed 1 mSv (100 mrem).

EPA has taken strong exception to the Nuclear Regulatory Commission standards for unrestricted release of contaminated sites (Luftig and Weinstock 1997; Trovato 1997). EPA believes that the standards are not adequately protective of human health and the environment in two important respects. First, the Nuclear Regulatory Commission's annual dose constraint of 0.25 mSv (25 mrem) for unrestricted release of contaminated sites does not comply with EPA's lifetime risk objective of 10^{-4}, which is applied in establishing preliminary remediation goals under CERCLA. EPA prefers a lower annual dose constraint of 0.15 mSv (15 mrem) to achieve the risk goal. Second, the Nuclear Regulatory Commission standards do not include a separate provision for groundwater protection in accordance with existing standards (MCLs) for public drinking-water supplies, and compliance with the Nuclear Regulatory Commission's annual dose constraint of 0.25 mSv (25 mrem) from all exposure pathways could result in radionuclide concentrations in groundwater in excess of drinking-water standards. The inclusion of such a provision would be in accordance with EPA's groundwater-protection strategy discussed above and with CERCLA and its implementing regulations, which specify that federal drinking-water standards are applicable or relevant and appropriate requirements (ARARs) for cleanup of groundwater (see chapter 7).

On the basis of the discussions in the previous two sections, the disagreement between EPA and the Nuclear Regulatory Commission over the adequacy of the Nuclear Regulatory Commission standards for unrestricted release of contaminated sites clearly is a matter of policy with no scientific or technical basis. The issue clearly is not whether the Nuclear Regulatory Commission standards protect human health and the environment because it is not the case that the resulting risks would be acceptable under EPA's approach but intolerable under the Nuclear Regulatory Commission's. The difference between an annual dose of 0.15 mSv and 0.25 mSv cannot reasonably be regarded as substantial, especially when the Nuclear Regulatory Commission also requires that the ALARA objective be applied in reducing doses below the specified dose constraint. Furthermore, the difference between an annual dose of 0.15 mSv and 0.25 mSv normally cannot be distinguished reliably in a dose assessment, given the substantial uncertainties in exposure pathway and dosimetry modeling. EPA's desire for a separate groundwater-protection requirement that complies with drinking-water standards also is not based on an a priori judgment about levels of contamination that are required for protection of public health without regard for the feasibility of achieving the standards. Thus, the disagreement between EPA and the Nuclear Regulatory Commission over appropriate cleanup standards for contaminated sites is entirely a matter of differences of opinion about reasonable approaches to risk management.

Differences in Guidances for Indoor Radon

The differences between EPA guidances for indoor radon and the recommendations of NCRP and ICRP are discussed in chapter 10. EPA's mitigation level for indoor radon is somewhat more restrictive than those recommended by NCRP and ICRP. This committee reiterates that these differences do not result from differences in the scientific and technical basis for the guidances. Rather, they result primarily from EPA's greater emphasis on reducing risks in the whole population on the basis of cost-benefit analysis, whereas the NCRP and ICRP guidances were based primarily on a concern for reducing exposures of the relatively few people who experience the highest risks.

Difference Between Environmental Protection Agency and National Council on Radiation Protection and Measurements Guidances for TENORM Other Than Indoor Radon

As discussed in chapter 10, EPA has issued proposed federal guidance on radiation protection of the public that includes an annual dose limit of 1 mSv (100 mrem) for all controlled sources combined, including human-made radionuclides and TENORM other than indoor radon (EPA 1994d). Furthermore, the proposed guidance specifies that the annual dose from individual sources or practices, including individual sources of exposure to TENORM other than indoor radon, should be limited to less than 1 mSv (100 mrem). In contrast, NCRP's recommended annual dose limit of 1 mSv (100 mrem) per year for members of the public (NCRP 1993a) does not apply to TENORM. Rather, the NCRP developed a separate recommendation that remedial actions be undertaken when the annual dose from exposure to natural sources only, including undisturbed natural background and TENORM other than indoor radon, exceeds 5 mSv (500 mrem). Therefore, although a direct comparison of the two guidances is not straightforward, the proposed EPA guidance, which applies to all sources of exposure to TENORM combined, should in most cases be considerably more restrictive than NCRP's recommended remedial-action level.

The difference between EPA and NCRP guidances for TENORM other than indoor radon does not result from differences in the scientific and technical basis of the guidances, in that both organizations assumed essentially the same risk related to radiation exposure for purposes of establishing the guidance. Rather, the difference results from differences in the approaches to risk management for TENORM. EPA regards TENORM other than indoor radon as a type of controlled source similar to sources of human-made radionuclides, so exposures to TENORM other than indoor radon are included in radiation-protection guidance that applies to human-made sources. However, NCRP regards TENORM as an enhanced form of natural background which should be treated separately from human-made sources for purposes of radiation protection. In addition, the difference between EPA's annual dose limit of 1 mSv (100 mrem) and NCRP's remedial action level of 5 mSv (500 mrem) reflects a difference in judgment about acceptable risks related to exposure to TENORM. Again, judgments about acceptable risk are strictly matters of policy.

Treatment of Natural Background in Establishing Guidances

Natural background radiation has played various roles in establishing guidances for control of exposure to radionuclides in the environment, depending primarily on whether or not the particular guidance applies to naturally occurring radionuclides (see chapters 5 and 7). EPA regulations that apply only to specific sources or practices involving human-made radionuclides generally do not take into account the magnitude and variability of natural background, because standards that were judged to provide an acceptable risk or to be reasonably achievable did not consider exposure to natural background. However, the annual dose limit of 1 mSv (100 mrem) for all controlled sources combined, including human-made radionuclides and TENORM other than indoor radon, in EPA's proposed federal guidance on radiation protection of the public (EPA 1994d), although it excludes exposures to natural background, was developed in recognition of the magnitude and variability of natural background (NCRP 1993a; ICRP 1991).

Natural background is important in developing guidances that apply to naturally occurring radionuclides. Current guidances for alpha-emitting radionuclides in drinking water, uranium and thorium mill tailings, and indoor radon are concerned only with naturally occurring radionuclides, and the development of guidances for these situations clearly required consideration of background levels of the radionuclides of concern. In the case of alpha-emitters in drinking water, the controllable exposures are due almost entirely to natural levels of radionuclides in groundwater or surface water; in the case of mill tailings and indoor radon, the levels of natural background provide a floor for any standards because the levels of the radionuclides of concern cannot be reduced below background.

Background also has been taken into account in different ways even for the same exposure situation involving naturally occurring radionuclides. A case in point involves guidances for indoor radon at uranium mill tailings sites. The initial EPA guidelines for homes built on sites contaminated with uranium mill tailings in Colorado specified remedial-action levels in excess of background (Harley 1996). When these guidelines were incorporated into EPA's uranium mill tailings standards in 40 CFR Part 192 (see chapter 7), the remedial-action level for indoor radon included background.

The issue of the most appropriate way of taking natural background into account in establishing guidances for radiation exposure is particularly important for TENORM other than indoor radon. As indicated by the discussions in the previous section, two approaches could be taken. Exposures to TENORM could be regulated without regard for the magnitude and variability of natural background, even though all radionuclides of concern are part of natural background. This approach is embodied, for example, in EPA's proposed federal guidance on radiation protection of the public (EPA 1994d) and the current guidance on cleanup of contaminated sites (Luftig and Weinstock 1997). Or, guidance could be developed for exposure to TENORM other than indoor radon and natural background combined; this is the approach recommended by NCRP (1993a).

Both approaches have advantages and disadvantages. The advantage of regulating without regard for the magnitude and variability of natural background is that controlled sources of exposure to TENORM would be regulated in the same way as human-made radionuclides; this would provide a desirable consistency in regulating all controlled sources. The disadvantage is that naturally occurring radionuclides resulting from human activities must be distinguishable from the undisturbed background of the same radionuclides. The distinction can be made if the difference between the levels of TENORM and natural background is sufficiently high, but the ability to measure TENORM with confidence depends on the magnitude and variability of background. Indeed, in some cases, it might be difficult to measure TENORM corresponding to low doses and risks, such as annual doses of 0.15 mSv (15 mrem) or lifetime risks of 10^{-4}. That disadvantage would probably be particularly important in establishing guidances for TENORM in soil, given the doses and risks associated with undisturbed natural background (see, for example, tables 2.8, 2.9 and 2.10).

Conversely, developing guidances for TENORM that include natural background has the disadvantage that controlled sources of TENORM would be regulated differently from human-made radionuclides. An advantage is that there would be no need to distinguish between TENORM and natural background; this could reduce the difficulties in verifying compliance with standards by means of environmental measurements.

Regardless of the approach used in taking natural background into account in developing guidances for TENORM, there is no fundamental scientific or technical basis for the choice. The choice would be based on risk-management policy and on considerations of the practicality of implementing the guidance, especially the ability to verify compliance by means of environmental measurements.

IMPLICATIONS OF GUIDANCES FOR RISK ASSESSMENT

The particular form that a guidance for TENORM might take has important implications for risk assessment, particularly with regard to issues that would need to be addressed in developing the guidance and issues that would be addressed in demonstrating compliance. That concern arises only with guidances for TENORM other than indoor radon, because of the availability of epidemiologic data that directly link concentrations of radon decay products in an exposure environment with increased risks of lung cancer.

This committee assumes that a guidance for TENORM other than indoor radon could be expressed in one of three ways: a limit on acceptable risk, a limit on acceptable dose, or limits on acceptable concentrations of radionuclides in various environmental media. Each has different implications for risk assessment.

If a guidance is expressed in terms of a limit on acceptable risk, all that is required in establishing the guidance, in principle, is a judgment about an acceptable risk for the exposure situations of concern. All issues for risk assessment could be addressed in demonstrating compliance with the limit. In practice, however, risk assessments normally would be used in developing guidances expressed only in terms of acceptable risk. For example, such assessments are required by the National Environmental Policy Act whenever a guidance would have substantial economic or environmental effects. In addition, some type of risk assessment normally would be needed to demonstrate that a proposed risk standard is reasonably achievable.

If a guidance is expressed in terms of a limit on acceptable dose that is based on the objective of achieving a particular risk, the one issue for risk assessment that would need to be addressed in developing the standard is the numerical value of the risk per unit dose. As indicated above, EPA normally uses the standard assumption for risk of 5×10^{-2} per sievert in establishing a dose standard based on a limit on acceptable risk. However, particularly in the case of TENORM, where only a few radionuclides are important, EPA could develop radionuclide-specific risk factors by using the methods discussed earlier, although this option would be attractive only if internal exposure to long-lived alpha-emitting radionuclides were more important than external exposure. With a dose standard, all other issues of risk assessment, particularly assessments of exposure pathways and the dose per unit exposure, would be considered in demonstrations of compliance.

Finally, if a guidance is expressed in terms of limits on acceptable concentrations of radionuclides in the environment, which are directly measurable, all issues of risk assessment—including exposure-pathway analysis, estimates of dose per unit exposure, and the approach to estimating risk—must be addressed in developing the standard, but none would need to be considered in demonstrations of compliance. This approach would allow the greatest opportunity for applying EPA's more rigorous methods of risk assessment discussed above in developing guidances for TENORM. However, it could be a considerable challenge to develop a standard expressed in terms of measurable quantities that reasonably could be applied to the variety of exposure situations of potential concern. Such complexity makes a standard expressed in terms of concentrations of radionuclides in the environment less attractive than a dose standard, which is the usual approach.

The particular form of guidances for TENORM that would be the most appropriate means of providing protection of human health and the environment is largely a matter of judgment, and there is no scientific or technical basis for the choice. The important concerns in choosing the particular form of any guidance include clarity of the regulatory approach, ease of implementation, and consistency with the approach used in other regulations, including those for human-made radionuclides.

12
Conclusions and Recommendations

This chapter presents a summary of the most important conclusions and recommendations developed from this study. The presentations in this chapter are organized as follows: the central conclusions and recommendations of the committee, including those which address the charge to the committee; the context for regulation of TENORM; risk-assessment issues underlying the regulation of TENORM; risk-management issues underlying regulations for TENORM; the comparability of guidances and regulations applicable to TENORM; and issues related to natural background radiation.

CENTRAL CONCLUSIONS AND RECOMMENDATIONS OF THE COMMITTEE

1. This committee finds that the differences between the Environmental Protection Agency (EPA) proposed and current guidelines for TENORM and similar guidelines developed by other organizations have no scientific or technical basis. There are some differences between various federal agencies in how they perform risk assessment, but the differences in their guidelines represent differences in policies for risk management.

2. This committee has not found a substantial body of relevant and appropriate scientific information that has not been used in the development of contemporary risk analysis for TENORM for purposes of developing and implementing guidelines. We emphasize that properties of TENORM do not differ from properties of other radionuclides in a way that would necessitate the development of different approaches to risk assessment. This committee has noted research needs related to improved understanding of the basis for high-

dose to low-dose extrapolation, particularly in the current use of the linear no-threshold model for cancer induction. We also call for research related to the improvement of exposure- and dose-assessment models through validation of parameters, for better standardization of measurement methods for TENORM, for better understanding of the effects of the chemical form and physical structure of TENORM on dose, and for greater insight into and documentation of the various uses and dispersal of TENORM.

3. The ALARA (as low as reasonably achievable) objective is the most important factor in guiding agency actions aimed at radiation protection—much more important than established regulatory limits or goals. To the extent that the ALARA objective is applied consistently to all exposure situations, all guidances and regulations would be consistent, provided that it is recognized that risks that are ALARA can vary considerably with the particular exposure situation.

CONTEXT FOR REGULATION OF TENORM

1. All natural media—earth, air, and water—and biota, including humans, contain naturally occurring radionuclides to some degree. Annual doses received by residents of the United States from all sources of natural radiation in the environment average about 3 mSv and are quite variable—estimated to range over a factor of about 4 for external sources and 20 for radon.

2. TENORM can be formed whenever NORM are moved from inaccessible locations to sites where there is a greater possibility of human exposure and whenever human activities process earth materials in a way that concentrates NORM. TENORM radionuclide concentrations and volumes vary greatly because of the diversity of sites, materials, and processes and because of the substantial variations in leachability, sorption, and biologic availability. Increases in radiation exposure from TENORM sources are typically local, rather than global, concerns.

RISK ASSESSMENT ISSUES UNDERLYING TENORM REGULATION

1. The committee notes that for radiation sources related to TENORM, including indoor radon, all regulatory and advisory groups have assumed about the same risk coefficients. This reflects a general acceptance by the scientific community of the linear no-threshold risk-extrapolation approach as a plausible and useful means of developing public-health regulations. The committee does

CONCLUSIONS AND RECOMMENDATIONS

not question the current suitability of the linear no-threshold model for regulatory purposes or the need for additional research as a basis for change in this model.

2. Exposure and dose or risk assessments used in developing standards should be reasonably realistic; that is, they should not be intended to greatly overestimate or underestimate actual effects for the exposure situations of concern.

3. For the purpose of developing guidelines, it is appropriate to develop stylized methods of exposure and dose or risk assessments for assumed reference conditions, provided that the assumed conditions are reasonably representative of the exposure situations of concern.

4. The chemical and physical forms of radionuclides in TENORM can greatly influence their environmental mobility and biologic availability. Exposure assessment for TENORM should consider such factors as bioavailability, leachability, and radon-emanation rates. Those factors are potentially important for developing guidelines for TENORM, and further research to understand them better should be undertaken.

5. Risk assessments for TENORM should also consider exposures to nonradioactive chemical agents that are often associated with TENORM.

RISK-MANAGEMENT ISSUES UNDERLYING TENORM GUIDELINES

1. All standards and guidances for radiation exposure are based fundamentally on judgments about the acceptability of health risks to the public or judgments about the achievability of health risks to the public. The latter, embodied in the ALARA principle, has been the most important consideration in controlling radiation exposures of the public for specific practices or sources, provided that the dose limit for all controlled sources is met.

2. Other considerations that may be important in developing guidances for radiation protection are the justification of practices (positive net benefit), the measurability of radioactivity in the environment at levels corresponding to the quantitative criteria in standards, and the magnitude and variability of natural background radiation and naturally occurring radionuclides in various environmental media.

3. The committee notes that neither EPA, which has primary responsibility for setting federal radiation standards, nor any other federal agency with responsibility for regulating radiation exposures has developed standards applicable to all exposure situations that involve TENORM. Instead, federal regulation of TENORM is fragmentary, and many potentially important sources of public exposure to TENORM are not regulated by any federal agency.

4. The committee strongly cautions against generalizing numerical guidance derived for a specific situation to another situation without sufficient thought as to the applicability to the new circumstance. For example, the soil-cleanup criteria developed under the Uranium Mill Tailings Radiation Control Act have been extended to many other situations by state and federal regulatory agencies, but many sources of TENORM have mineralogic characteristics and processing histories that differ greatly from those of uranium mill tailings, and therefore have different radon-emanation coefficients, leachability, and bioavailability.

COMPARABILITY OF GUIDANCES AND REGULATIONS POTENTIALLY RELATED TO TENORM

1. The implied health risks for the different radiation guidances and regulations potentially applicable to TENORM vary over several orders of magnitude.

2. Although consistency among the many guidances and regulations for radionuclides is desirable, there are valid reasons not to expect it, including agency differences in statutory and judicial mandates for standards, in the regulatory bases of standards, in the applicability of standards, in the population groups of primary concern to the standards, and in the considerations of natural background in setting standards. Furthermore, the various guidances for TENORM were developed at different times, and the basic assumptions about radiation risk have changed.

3. The committee concludes that different guidances and regulations should not be compared unless their bases and their applicability are well understood and the quantitative criteria are interpreted properly. Otherwise, misleading conclusions about the meaning and importance of differences in implied risks might result.

CONCLUSIONS AND RECOMMENDATIONS

4. We conclude that the large differences in implied health risks among the various guidelines and regulations do not necessarily mean that the different standards are inconsistent with regard to the determination of an acceptable risk to the public. The principle that exposures should be maintained ALARA, economic and social factors being taken into account, appears to be the most important factor in determining risks actually experienced for any controllable exposure situation.

5. The more stringent mitigation levels for indoor radon recommended by EPA, compared with those of most other countries and the National Council on Radiation Protection and Measurements (NCRP) and the International Commission on Radiological Protection (ICRP), do not result from differences in scientific opinion about risks posed by exposure to indoor radon. Rather, they result primarily from such factors as differences in average radon levels in homes, differences in judgments based on cost-benefit analysis about levels of radon that are reasonably achievable after mitigation, and differences in whether the guidances focus primarily on reduction of risks to individuals receiving the highest exposures or on reduction of population risks.

6. In most cases, EPA's current guidances on acceptable exposures to TENORM other than indoor radon also are somewhat more restrictive than the guidances developed by some of the states, the NCRP, and the Health Physics Society. However, direct comparisons of these guidances are difficult and potentially misleading, because of such factors as differences in whether exposures to natural background are included, the difference between a regulatory limit and a goal, and the use of a dose criterion in some cases but activity concentrations of particular radionuclides in others.

7. The committee does not view the current differences in how the agencies develop and carry out their recommendations, although perhaps confusing, as necessarily resulting in important differences in protection of public health. However, the committee does caution that, as the regulations are developed and acted on, continued attention to the factors that affect radiation dose and risk for specific TENORM situations is crucial for consistently protective, cost-effective radiation control. In addition, further study on issues of cost-benefit and other nonscientific concerns could be important in regulating TENORM, given the magnitude and variability of natural background.

8. The committee has considered the disagreement between EPA and the Nuclear Regulatory Commission over the adequacy of the Commission's standards for unrestricted release of contaminated sites. The committee believes that the disagreement is a matter of policy with no scientific or technical basis.

The difference between EPA's proposed annual dose constraint of 0.15 mSv and the Commission's dose constraint of 0.25 mSv cannot reasonably be regarded as significant, particularly when the Commission also requires that the ALARA objective be applied in reducing doses below 0.25 mSv. Furthermore, the difference between 0.15 mSv and 0.25 mSv normally cannot be distinguished reliably in a dose assessment. The disagreement over the need for a separate groundwater-protection standard consistent with existing standards for radioactivity in drinking water also is a matter of differences in policies for risk management.

9. EPA and the Nuclear Regulatory Commission have worked together to produce a valuable document on risk harmonization that effectively summarizes their similarities and differences in the approach to radiation protection. The committee commends the agencies for having done so and recommends that they pursue this approach further. The committee recognizes the different objectives and histories of the two agencies, but it is good public policy to reconcile the existing differences in approaches to risk management with an eye to better, more timely, and more efficient compliance activities by the regulated community, and greater acceptance by Congress and the public.

ISSUES RELATED TO BACKGROUND RADIATION

1. The committee concludes that background radiation levels of NORM are highly relevant to regulation of TENORM because the radionuclides being regulated as TENORM are identical with those in nature. Arguments concerning small differences in the target regulatory level at small fractions of the natural background tend to pale into insignificance in comparison with natural background levels and their local and regional variations.

2. Considering only external photon exposure, the committee notes that EPA's proposed 0.15-mSv (15-mrem) standard is equivalent to an incremental increase in the concentration of radium-226 in soil of about the usual natural background level of 0.04 Bq/g (1 pCi/g). In view of the ubiquitousness of ^{226}Ra in soil and the substantial local variation in natural background, it is likely to be difficult to implement a 0.15-mSv (15-mrem) soil-cleanup standard for radium, particularly when the contamination is only marginally above the local background. That is especially the case if potential exposures to indoor radon are included in complying with the standard.

3. As a practical matter, the implications of existing levels and variability of natural radionuclide concentrations and doses received by humans should receive careful consideration in the regulation of TENORM.

REFERENCES

Advisory Committee on X-ray and Radium Protection. 1941. Safe Handling of Radioactive Luminous Compounds. In: National Bureau of Standards Handbook 27. Washington, DC: National Bureau of Standards.

Ahmed JU. 1992. Regulatory approach toward controlling exposure to radon in dwellings. Radiation Protection Dosimetry 45(1/4):745-750.

Alter HW, Oswald RA. 1987. Nationwide distribution of indoor radon measurements: a preliminary data base. Journal of the Air Pollution Control Association 37:227-231.

Anonymous. 1995. New rules could bring ceramics industry under LLW standards. Nuclear Waste News 15(44):435.

Baes CF, Marland G. 1989. Evaluation of Cleanup Levels for Remedial Action at CERCLA Sites Based on a Review of EPA Records of Decision. ORNL-6479. Oak Ridge, TN: Oak Ridge National Laboratory.

Bair WJ, Sinclair WK. 1992. Response to Drs. Puskin and Nelson note. Health Physics 63(5):590.

Barretto PMC, Clark RB, Adams JAS. 1975. Physical characteristics of radon-222 emanation from rocks, soils, and minerals: its relation to temperature and alpha dose. In: Adams JAS, Lowder WM, Gesell TF (eds.), The Natural Radiation Environment II, August 7-11 1972, Houston, TX. CONF-720805-P1. Washington, DC: Energy Research and Development Administration. pp. 731-740.

Baxter MS. 1996. Technologically enhanced radioactivity: an overview. Journal of Environmental Radioactivity 32(1-2):3-17.

Beck HL. 1966. Environmental gamma radiation from deposited fission products, 1960-1964. Health Physics 12:313-322.

Beck HL. 1975. The physics of environmental radiation fields. In: Adams JAS, Lowder WM, Gesell TF (eds.), The Natural Radiation Environment II. CONF-720805-P1. Washington, DC: Energy Research and Development Administration. pp. 101-133.

Beck HL. 1980. Exposure Rate Conversion Factors for Radionuclides Deposited on the Ground. EML-378. New York: Department of Energy.

Bernhardt DE, Owen DH, Rogers VC. 1996. Assessments of NORM in pipe from oil and gas production. In: Mossman KL, Thiemann KB (eds.), NORM/NARM: Regulation and Risk Assessment, Proceedings of the 29th Midyear Topical Meeting of the Health Physics Society. McLean, VA: Health Physics Society. pp. 133-138.

Bliss JD. 1978. Radioactivity in Selected Mineral Extraction Industries: A Literature Review. Technical Note ORP/LV-79-1. Washington, DC: Environmental Protection Agency.

Blot WJ, Xu ZY, Boice JD, Zhao DZ, Stone BJ, Sun J. 1990. Indoor radon and lung cancer in China. Journal of the National Cancer Institute 82(12):1025-1030.

Boothe GF, Stewart-Smith D, Wagstaff D, Dibblee M. 1980. The radiological aspects of zircon sand use. Health Physics 38(3):393-398.

Boyle RW. 1982. Geochemical Prospecting for Thorium and Uranium Deposits. New York, NY: Elsevier.

Breas GM, van der Vaart PI. 1997. Scrapmetals and NORM. In: Proceedings of the International Symposium on Radiological Problems with Natural Radioactivity in the Non-Nuclear Industry September 8-10 1997. Arhem, The Netherlands: KEMA. Section 6.5.

Brown SL. 1992. Harmonizing chemical and radiation risk management. Environmental Science & Technology 26(12):2336-2338.

Cameron FX. 1996. The odyssey of the good ship NORM: the search for a regulatory safe harbor. In: Mossman KL, Thiemann KB (eds.), NORM/NARM: Regulation and Risk Assessment, Proceedings of the 29th Midyear Topical Meeting of the Health Physics Society. McLean, VA: Health Physics Society. pp. 13-18.

Carter MW, Coundouris AN. 1993. Radiation protection in the mineral sands industry in New South Wales. Radiation Protection in Australia 11:90-96.

Chhabra AS. 1966. Radium-226 in food and man in Bombay and Kerala State (India). British Journal of Radiology 39(458):141-146.

Clarke RH. 1995. ICRP recommendations applicable to the mining and minerals processing industries and to natural sources. Health Physics 69(4):454-460.

Clarke RH. 1996. Control of exposure to natural radiation: an ICRP perspective. Environment International 22(Suppl. 1):S105-S110.

Clay DR. 1991. Role of the Baseline Risk Assessment in Superfund Remedy Selection Decisions. Directive 9355.0-30. Washington, DC: Environmental Protection Agency, Office of Solid Waste and Emergency Response.

Cohen BL. 1986. A national survey of Rn-222 in U.S. homes and correlating factors. Health Physics 51(2):175-183.

Cohen BL. 1989. Measured radon levels in U.S. homes. In: Proceedings of the Twenty-Fourth Annual Meeting of the National Council on Radiation Protection and Measurements. NCRP Proceedings No. 10. Bethesda, MD: National Council on Radiation Protection and Measurements. pp. 170-181.

Cohen BL. 1991. Variation of radon levels in U.S. homes correlated with house characteristics, location, and socioeconomic factors. Health Physics 60(5):631-642.

Cohen BL, Kulwicki DR, Warner Jr KR, Grassi CL. 1984. Radon concentrations inside public and commercial buildings in the Pittsburgh area. Health Physics 47(3):399-405.

Colgan PA, Gutierrez J. 1996. National approaches to controlling exposure to radon. Environment International 22(Suppl. 1):S1083-S1092.

Coll L. 1997. Chemical Decontamination of NORM Scales. Research Triangle Park, NC: CORPEX Technologies Inc.

Cothern CR, Lappenbusch WL. 1984. Compliance data for the occurrence of radium and gross -particle activity in drinking water supplies in the United States. Health Physics 46(3):503-510.

CRCPD. (Conference of Radiation Control Program Directors, Inc.). 1978. Natural Radioactivity Contamination Problems: A Report of the Task Force. EPA 520/4-77-015. Washington, DC: Environmental Protection Agency.

CRCPD. (Conference of Radiation Control Program Directors, Inc.). 1997. Part N—Regulation and Licensing of Technologically Enchanced Naturally Occurring Radioactive Materials (TENORM) Draft. Frankfort, KY: Conference of Radiation Control Program Directors (February).

CRCPD (Conference of Radiation Control Program Directors, Inc.). 1998. Available at: http://www.crcpd.org/.

Cross FT, Harley NH, Hofmann W. 1985. Health effects and risks from ^{222}Rn in drinking water. Health Physics 48(5):649-670.

Cullen TL, Paschoa AS. 1978. Radioactivity in certain products in Brazil. In: Moghissi AA, Parras P, Carter MW, Barkers RF (eds.), Radioactivity in Consumer Products. NUREG/CP-0001. Washington, DC: Nuclear Regulatory Commission. pp. 376-379.

Curtis SB, Dye DL, Sheldon WR. 1966. Hazard from highly ionizing radiation in space. Health Physics 12:1069-1075.

DHHS. (Department of Health and Human Services). 1989. Reducing the Health Consequences of Smoking: 25 years of Progress. A Report of the Surgeon General. 89-8411. Washington, DC: Department of Health and Human Services.

DOE. (Department of Energy). 1988. Radioactive Waste Management. DOE Order 5820.2A. Washington, DC: Department of Energy.

DOE. (Department of Energy). 1990. Radiation Protection of the Public and the Environment. DOE Order 5400.5. Washington, DC: Department of Energy.

DOE. (Department of Energy). 1993a. 10 CFR Part 834—Radiation protection of the public and the environment. Proposed rule. Federal Register 58(56):16268-16322.

DOE. (Department of Energy). 1993b. Generic Protocol for Th-230 Cleanup/Verification at UMTRA Project Sites. Albuquerque, NM: Department of Energy.

DOE. (Department of Energy). 1996. Integrated Data Base Report—1995: U.S. Spent Nuclear Fuel and Radioactive Waste Inventories, Projections, and Characteristics. DOE/RW-0006, Rev. 12. Washington, DC: Department of Energy.

DOE (Department of Energy). 1997. Reindustrialization, Reuse, and Recycling on FUSRAP Environmental Remediation Projects. Available at: http://www.fusrap.doe.gov/techpap/reindust.html.

Dunning Jr DE, Leggett RW, Sullivan RE. 1984. An assessment of health risk from radiation exposures. Health Physics 46(15):1035-1051.

Dunning Jr DE, Leggett RW, Yalcintas MG. 1980. A Combined Methodology for Estimating Dose Rates and Health Effects from Exposure to Radioactive Pollutants. ORNL/TM-7105. Oak Ridge, TN: Oak Ridge National Laboratory.

Durrance EM. 1986. Radioactivity in Geology: Principles and Applications. New York, NY: John Wiley and Sons.

EC. (European Communities). 1990. Commission recommendation of 21 February 1990 on the protection of the public against indoor exposure to radon. Official Journal of the European Communities L80:26-28.

Eckerman KF, Leggett RW, Nelson CB, Puskin JS, Richardson ACB. 1998. Health Risks from Low-Level Environmental Exposure to Radionuclides, Federal Guidance Report No. 13, Part 1-Interim Version. EPA 402-R-97-014. Oak Ridge National Laboratory and Environmental Protection Agency.

Eckerman KF, Ryman JC. 1993. External Exposure to Radionuclides in Air, Water, and Soil, Federal Guidance Report No. 12. EPA 402-R-93-081. Washington, DC: Oak Ridge National Laboratory and Environmental Protection Agency.

Eckerman KF, Wolbarst AB, Richardson ACB. 1988. Limiting Values of Radionuclide Intake and Air Concentration and Dose Conversion Factors for Inhalation, Submersion, and Ingestion, Federal Guidance Report No. 11. EPA-520/1-88-020. Washington, DC: Oak Ridge National Laboratory and Environmental Protection Agency.

Eisenbud M, Gesell T. 1997. Environmental Radioactivity. 4th edition. San Diego, CA: Academic Press.

EPA. (Environmental Protection Agency). 1977. Proposed Guidance on Dose Limits for Persons Exposed to Transuranium Elements in the Natural Environment. EPA/520/4-77/016. Washington, DC: Environmental Protection Agency.

EPA. (Environmental Protection Agency). 1982. Final Environmental Impact Statement for Remedial Action Standards for Inactive Uranium Processing Sites (40 CFR 192). EPA 520/4/82/013-1. Washington, DC: Environmental Protection Agency.

EPA. (Environmental Protection Agency). 1985. 40 CFR Part 191—Environmental standards for the management and disposal of spent nuclear fuel, high-level and transuranic radioactive wastes. Final Rule. Federal Register 50(182):38066-38089.

EPA. (Environmental Protection Agency). 1987a. Radiation protection guidance to federal agencies for occupational exposure. Federal Register 52(17):2822-2834.

EPA. (Environmental Protection Agency). 1987b. Radon Reference Manual. EPA 520/1-87/20. Washington, DC: Environmental Protection Agency.

EPA. (Environmental Protection Agency). 1989a. 40 CFR Part 764—Environmental standards for management, storage, and land disposal of naturally occurring and accelerator-produced radioactive waste. Draft proposed rule (April 6).

EPA. (Environmental Protection Agency). 1989b. Environmental Impact Statement for Proposed NESHAPs for Radionuclides. 520/1-89-005. Washington, DC: Environmental Protection Agency.

EPA. (Environmental Protection Agency). 1989c. Human Health Evaluation Manual (Part A). Risk Assessment Guidance for Superfund. EPA/540/1-89/002. Washington, DC: Environmental Protection Agency.

EPA. (Environmental Protection Agency). 1989d. 40 CFR Part 61—National emission standards for hazardous air pollutants; radionuclides. Final rule. Federal Register 54(240):51654-51715.

EPA. (Environmental Protection Agency). 1990. Suggested Guidelines for the Disposal of Drinking Water Treatment Wastes Containing Naturally Occurring Radionuclides (Draft). Washington, DC: Environmental Protection Agency, Office of Drinking Water.

EPA. (Environmental Protection Agency). 1991a. 40 CFR Parts 141, 142—National primary drinking water regulations; radionuclides, Proposed rule. Federal Register 56(138):33050-33127.

EPA. (Environmental Protection Agency). 1991b. Protecting the Nation's Ground Water: EPA's Strategy for the 1990s. Washington, DC: Environmental Protection Agency.

EPA. (Environmental Protection Agency). 1992a. Manual of Protective Action Guides and Protective Actions for Nuclear Incidents. EPA 400-R-92-001. Washington, DC: Environmental Protection Agency.

EPA. (Environmental Protection Agency). 1992b. National Residential Radon Survey. EPA 402-R-92-011. Washington, DC: Environmental Protection Agency.

EPA. (Environmental Protection Agency). 1992c. Technical Support Document for the 1992 Citizens Guide to Radon. EPA 400-R-92-011. Washington, DC: Environmental Protection Agency, Office of Radiation and Indoor Air.

EPA. (Environmental Protection Agency). 1992d. 40 CFR Part 61—National emissions standards for hazardous air pollutants; national emissions standards for radon emissions from phosphogypsum stacks. Final rule. Federal Register 57(107):23305-23320.

EPA. (Environmental Protection Agency). 1993a. Basis for the soil cleanup criteria in 40 CFR Part 192, Memorandum from Oge, M. Washington, DC: US Environmental Protection Agency, Office of Radiation and Indoor Air (June 10).

EPA. (Environmental Protection Agency). 1993b. Diffuse NORM Wastes: Waste Characterization and Preliminary Risk Assessment. RAE-9232/1-2, Draft report. Washington, DC: Environmental Protection Agency.

EPA. (Environmental Protection Agency). 1994a. 40 CFR Part 193—Environmental radiation protection standards for the management, storage and disposal of low-level radioactive waste. Draft proposed rule. Washington, DC: Environmental Protection Agency (November 30).

EPA. (Environmental Protection Agency). 1994b. 40 CFR Part 196—Environmental Protection Agency radiation site cleanup regulation. Draft proposed rule. Washington, DC: Environmental Protection Agency (May 11).

EPA. (Environmental Protection Agency). 1994c. Estimating Radiogenic Cancer Risks. EPA 402-R-93-076. Washington, DC: Environmental Protection Agency.

EPA. (Environmental Protection Agency). 1994d. Federal radiation protection guidance for exposure of the general public, Proposed recommendations. Federal Register 59(246):66414-66428.

EPA. (Environmental Protection Agency). 1994e. Suggested Guidelines for Disposal of Drinking Water Treatment Wastes Containing Radioactivity. Washington, DC: Environmental Protection Agency, Office of Ground Water and Drinking Water.

EPA. (Environmental Protection Agency). 1995. 40 CFR Part 192—Groundwater standards for remedial actions at inactive uranium processing sites, Final rule. Federal Register 60(7):2854-2871.

EPA. (Environmental Protection Agency). 1997. 40 CFR Parts 141 and 142—Withdrawal of proposed national primary drinking water regulations for radon-222. Federal Register 62(15):42221-42222.

EPA and DHHS. (Environmental Protection Agency and Department of Health and Human Services). 1986. A Citizen's Guide to Radon. OPA-86-004. Washington, DC: Environmental Protection Agency.

EPA and DHHS. (Environmental Protection Agency and Department of Health and Human Services). 1994. A Citizen's Guide to Radon. EPA ANR-464, DHHS 402-K92-001. Washington, DC: U.S. Government Printing Office.

EPA-SAB. (Environmental Protection Agency-Science Advisory Board). 1992. Commentary on Harmonizing Chemical and Radiation Risk Reduction Strategies. EPA-SAB-RAC-COM-92-007. Washington, DC: Environmental Protection Agency.

Euratom. 1996. Council Directive 96/29/Euratom of 13 May 1996. Official Journal of the European Communities L159:39.

Faure G. 1986. Principles of Isotope Geology. 2nd edition. New York: Wiley.

Fisenne IM. 1993. Initial study of Pb-210 in indoor air. Health Physics 64(4):423-425.

Fisenne IM, Keller HW. 1970. Radium-226 in the diet of two U.S. cities. HASL-224. New York: Atomic Energy Commission.

Fisenne IM, Keller HW, Harley NH. 1981. Worldwide measurement of ^{226}Ra in human bone: estimate of skeletal alpha dose. Health Physics 40(2):163-171.

Fowler WA. 1967. Nuclear Astrophysics. Philadelphia, PA: American Philosophical Society.

FRC. (Federal Radiation Council). 1960. Radiation protection guidance for federal agencies. Federal Register 25(97):4402-4403.

FRC. (Federal Radiation Council). 1961. Radiation protection guidance for federal agencies. Federal Register 26(185):9057-9058.

GAO. (Government Accounting Office). 1994. Nuclear Health and Safety: Concensus on Acceptable Radiation Risk to the Public is Lacking. GAO/RCED-94-190. Washington, DC: Government Accounting Office.

Gesell TF. 1975. Occupational radiation exposure due to ^{222}Rn in natural gas and natural gas products. Health Physics 29:681-687.

Gesell TF. 1983. Background atmospheric ^{222}Rn concentrations outdoors and indoors: a review. Health Physics 45(2):289-302.

Gesell TF, Adams JAS. 1975. Geothermal power plants: environmental impact. Science 189(4199):328.

Gesell TF, Prichard HM. 1975. The technologically enhanced natural radiation environment. Health Physics 28:361-366.

Gilkeson RH, Perry EC, Cowart JB, Holtzman RB. 1984. Isotope Studies of the Natural Sources of Radium in Groundwater in Illinois. UILU-WRC-84-187. Urbana, IL: University of Illinois.

Gold S, Barkhau HW, Shleien B, Kahn B. 1964. Measurement of naturally occurring radionuclides in air. In: Adams JAS, Lowder WM (eds.), The Natural Radiation Environment. Chicago, IL: University of Chicago Press. pp. 369-382.

Guimond RJ. 1978. The radiological aspects of fertilizer utilization. In: Moghissi AA, Paras P, Carter MW Barker RF (eds.), Radioactivity in Consumer Products. NUREG/CP0003. Washington, DC: Nuclear Regulatory Commission. pp. 380-393.

Hahn O. 1936. Applied Radiochemistry. Ithaca, NY: Cornell University Press.

Harada K, Burnett WC, LaRock PA. 1989. Polonium in Florida groundwater and its possible relationship to the sulfur cycle and bacteria. Geochimica et Cosmochimica Acta 53(1):143-150.

Harley NH. 1996. Radon: over- or under-regulated? In: Mossman KL, Thiemann KB (eds.), NORM/NARM: Regulation and Risk Assessment, Proceedings of the 29th Midyear Topical Meeting of the Health Physics Society. McLean, VA: Health Physics Society. pp. 39-45.

Heaton B, Lambley J. 1995. TENORM in oil, gas and mineral mining. Applied Radiation and Isotopes 46(6/7):577-581.

Hess CT, Michel J, Horton TR, Prichard HM, Coniglio WA. 1985. The occurrence of radioactivity in public water supplies in the United States. Health Physics 48(5):553-586.

Hess CT, Norton SA, Brutsaert WF, Lowry JF, Weiffenbach CV, Casparius RE, Coombs EG, Brandow JE. 1981. Investigation of ^{222}Rn, ^{226}Ra, and U in Air and Groundwaters of Maine. B017-ME. University of Maine at Orono: Land and Water Resources Center.

Horton TR. 1985. Nationwide Occurrence of Radon and Other Natural Radioactivity in Public Water Supplies. EPA 520/5-85-008. Washington, DC: Environmental Protection Agency.

HPS. (Health Physics Society). 1992. Position Paper on Radiation Dose Limits for the General Public. McLean, VA: Health Physics Society.

HPS. (Health Physics Society). 1993. Radiation standards for site cleanup and restoration. Health Physics Society Newsletter 21(6):7-10.

Hull CD, Burnett WC. 1996. ^{238}U decay-series nuclides in fluids within a Florida phosphogypsum storage stack. In: Mossman KL, Thiemann KB (eds.), NORM/NARM: Regulation and Risk Assessment, Proceedings of the 29th Midyear Topical Meeting of the Health Physics Society. McLean, VA: Health Physics Society. pp. 111-120.

Hultqvist B. 1956. Studies on naturally occuring ionizing radiation. Kgl. Sv. Vetenskapsakade Handl. 4(Suppl).

IAEA. (International Atomic Energy Agency). 1982. Basic Safety Standards for Radiation Protection. Vienna, Austria: International Atomic Energy Agency.

IAEA. (International Atomic Energy Agency). 1988. Principles for the Exemption of Radiation Sources and Practices from Regulatory Control. Safety Series No. 89. Vienna, Austria: International Atomic Energy Agency.

IAEA. (International Atomic Energy Agency). 1990. The Environmental Behaviour of Radium. Technical Report Series No. 310. Vienna, Austria: International Atomic Energy Agency.

IAEA. (International Atomic Energy Agency). 1991. Airborne Gamma-Ray Spectrometer Surveying. Technical Report Series No. 323. Vienna: IAEA.

IAEA. (International Atomic Energy Agency). 1992. Effects of Ionizing Radiation on Plants and Animals at Levels Implied by Current Radiation Protection Standards. Technical Report Series No. 332. Vienna, Austria: IAEA.

IAEA. (International Atomic Energy Agency). 1994. Handbook of Parameter Values for the Prediction of Radionuclide Transfer in Temperate Environments. Technical Report Series No. 364. Vienna, Austria: International Atomic Energy Agency.

IAEA. (International Atomic Energy Agency). 1995. The Principles of Radioactive Waste Management. Safety Series No. 111-F. Vienna, Austria: International Atomic Energy Agency.

IAEA. (International Atomic Energy Agency). 1996a. International Basic Safety Standards for Protection Against Ionizing Radiation and for the Safety of Radiation Sources. Safety Series No. 115-F. Vienna, Austria: International Atomic Energy Agency.

IAEA. (International Atomic Energy Agency). 1996b. Modelling of Radionuclide Interception and Loss Processess in Vegetation and of Transfer in Semi-Natural Ecosystems, Second Report of VAMP Terrestrial Working Group. IAEA-TECDOC-857. Vienna, Austria: International Atomic Energy Agency.

IARC. (International Agency for Research on Cancer). 1988. IARC Monographs on the Evaluation of Carcinogenic Risks to Humans, Vol. 43: Man-made Mineral Fibres and Radon. Lyon, France: International Agency for Research on Cancer.

Ibrahim SA, Wrenn ME, Singh NP, Cohen N, Saccomano G. 1983. Thorium concentrations in human tissues from two U.S. populations. Health Physics 44(Suppl. 1):213-220.

ICRP. (International Commission on Radiological Protection). 1977. Recommendations of the International Commission on Radiological Protection. ICRP Publication 26, Annals of the ICRP 1(3). Oxford: Pergamon Press.

ICRP. (International Commission on Radiological Protection). 1979. Limits for Intakes of Radionuclides by Workers. ICRP Publication 30, Part 1, Annals of the ICRP 2(3/4). Oxford: Pergamon Press.

ICRP. (International Commission on Radiological Protection). 1981. Limits for Inhalation of Radon Daughters by Workers. ICRP Publication 32, Annals of the ICRP 6(1). Oxford: Pergamon Press.

ICRP. (International Commission on Radiological Protection). 1984. Principles for Limiting Exposure of the Public to Natural Sources of Radiation. ICRP Publication 39, Annals of the ICRP 14(1). Oxford: Pergamon Press.

ICRP. (International Commission on Radiological Protection). 1986. Radiation Protection of Workers in Mines. ICRP Publication 47, Annals of the ICRP 16(1). Oxford: Pergamon Press.

ICRP. (International Commission on Radiological Protection). 1987a. Data for Use in Protection Against External Radiation. ICRP Publication 51, Annals of the ICRP 17(2/3). Oxford: Pergamon Press.

ICRP. (International Commission on Radiological Protection). 1987b. Lung Cancer Risk from Indoor Exposures to Radon Daughters. ICRP Publication 50, Annals of the ICRP 17(1). Oxford: Pergamon Press.

ICRP. (International Commission on Radiological Protection). 1989a. Age-Dependent Doses to Members of the Public from Intake of Radionuclides, Part. 1. ICRP Publication 56, Annals of the ICRP 20(2). Oxford: Pergamon Press.

ICRP. (International Commission on Radiological Protection). 1989b. Optimization and Decision-Making in Radiological Protection. ICRP Publication 55, Annals of the ICRP 20(1). Oxford: Pergamon Press.

ICRP. (International Commission on Radiological Protection). 1991. 1990 recommendations of the International Commission on Radiological Protection. ICRP Publication 60, Annals of the ICRP 21(1-3). Oxford: Pergamon Press.

ICRP. (International Commission on Radiological Protection). 1993a. Age-Dependent Doses to Members of the Public from Intake of Radionuclides: Part 2, Ingestion Dose Coefficients. ICRP Publication 67, Annals of the ICRP 23(3/4). Oxford: Pergamon Press.

ICRP. (International Commission on Radiological Protection). 1993b. Protection Against Radon-222 at Home and at Work. ICRP Publication 65, Annals of the ICRP 23(2). Oxford: Pergamon Press.

ICRP. (International Commission on Radiological Protection). 1994. Human Respiratory Tract Model for Radiological Protection. ICRP Publication 66, Annals of the ICRP 24(1/3). Oxford: Pergamon Press.

ICRP. (International Commission on Radiological Protection). 1995. Age-Dependent Doses to Members of the Public from Intake of Radionuclides: Part. 3, Ingestion Dose Coefficients. ICRP Publication 69, Annals of the ICRP 23(3/4). Oxford: Pergamon Press.

ICRP. (International Commission on Radiological Protection). 1996. Age-Dependent Doses to Members of the Public from Intake of Radionuclides: Part 4, Inhalation Dose Coefficients. ICRP Publication 71, Annals of the ICRP 25(3/4). Oxford: Pergamon Press.

Janssens A. 1992. CEC radon control policy recommendations. Radiation Protection Dosimetry 45(1/4):759-761.

Janssens A, Markkanen M. 1997. Provisions on natural radiation in the new basic safety standards directive. In: Proceedings of the International Symposium on Radiological Problems with Natural Radioactivity in the Non-Nuclear Industry, September 8-10, 1997. Arnhem, The Netherlands: KEMA. Section 1.1.

Jones MW, Dron EM. 1992. Domestic radon reduction strategy in the United Kingdom. Radiation Protection Dosimetry 45(1/4):763-766.

Kauranen P, Miettinen JK. 1969. Po-210 and Pb-210 in the arctic food chain and the natural radiation exposure of Lapps. Health Physics 16:287-296.

Klemic G. 1996. Environmental radiation monitoring in the context of regulations on dose limits to the public. In: Proceedings of 1996 Congress of the International Radiation Protection Association, April 14-19, 1996. Vienna, Austria: Austrian Association for Radiation Protection. pp. 321-328.

Kocher DC. 1988. Review of radiation protection and environmental radiation standards for the public. Nuclear Safety 29(4):463-475.

Kocher DC. 1989. Relationship between kidney burden and radiation dose from chronic ingestion of U: implications for radiation standards for the public. Health Physics 57(1):9-15.

Kocher DC, Hoffman FO. 1991. Regulating environmental carcinogens: where do we draw the line? Environmental Science & Technology 25(12):1986-1989.

Kocher DC, Hoffman FO. 1992. Reply to Weisburger regarding 'regulating environmental carcinogens.' Environmental Science & Technology 26(5):845-846.

Kocher DC, O'Donnell FR. 1987. Considerations on a De Minimis Dose and Disposal of Exempt Concentrations on Radioactive Wastes. ORNL/TM-10388. Oak Ridge, TN: Oak Ridge National Laboratory.

Kolb W, Wojik M. 1985. Enhanced radioactivity due to natural oil and gas production and related radiological problems. Science of the Total Environment 45:77-84.

Kraemer TF, Curwick PB. 1991. Radium isotopes in the lower Mississippi River. Journal of Geophysical Research 96(C2):2797-2806.

Lancée P, Eylander JGR, Hartog FA, Jonkers G. 1997. Review on E&P NORM volume reduction techniques and waste disposal routes. In: Proceedings of the International Symposium on Radiological Problems with Natural Radioactivity in the Non-nuclear Industry, September 8-10, 1997. Arnhem, The Netherlands: KEMA. Section 6.2.

Landa ER. 1980. Isolation of Uranium Mill Tailings and Their Component Radionuclides from the Biosphere—Some Earth Science Perspectives. Geological Survey Circular 814. Reston, VA: U.S. Geological Survey.

Landa ER. 1982. Leaching of radionuclides from uranium ore and mill tailings. Uranium 1:53-64.

Landa ER. 1984. Geochemical and radiological characterization of soils from former radium processing sites. Health Physics 46(2):385-394.

Landa ER. 1987. Radium-226 contents and Rn emanation coefficients of particle-size fraction of alkaline, acid and mixed U mill tailings. Health Physics 52(3):303-310.

Landa ER, Reid DF. 1983. Sorption of ^{226}Ra from oil-production brine by sediments and soils. Environmental Geology 5(1):1-8.

Landa ER, Stieff LR, Germani MS, Tanner AB, Evans JR. 1994. Intense alpha-particle emitting crystallites in uranium mill wastes. Nuclear Geophysics 8(5):443-454.

Langroo MK, Wise KN, Duggleby JC, Kotler LH. 1991. A nationwide survey of Rn-222 and gamma radiation levels in Australian homes. Health Physics 61(6):753-761.

Lawrence EP, Wanty RB, Nyberg P. 1992. Contribution of Rn-222 in domestic water supplies to Rn-222 in indoor air in Colorado homes. Health Physics 62(2):171-177.

Lee DW, Kocher DC, Wang JC. 1996. Operating limit evaluation for disposal of uranium enrichment plant wastes. In: Mossman KL, Thiemann KB (eds.), NORM/NARM: Regulation and Risk Assessment, Proceedings of the 29th Midyear Topical Meeting of the Health Physics Society. McLean, VA: Health Physics Society. pp. 91-98.

Lee DW, Wang JC, Kocher DC. 1995. Operating Limit Study for the Proposed Solid Waste Landfill at Paducah Gaseous Diffusion Plant. ORNL/TM-13008. Oak Ridge, TN: Oak Ridge National Laboratory.

Leggett RW. 1989. The behavior and chemical toxicity of U in the kidney: a reassessment. Health Physics 57(3):365-383.

Lettner H, Steinhäusler F. 1988. Radon exhalation of waste gypsum recycled as building material. Radiation Protection Dosimetry 24(1-4):415-417.

Linsalata P. 1994. Uranium and thorium decay series radionuclides in human and animal foodchains—a review. Journal of Environmental Quality 23(4):633-642.

Lowry JD. 1983. Removal of Radon from Water Using Granular Activated Carbon Adsorption. 14-34-0001-2121, A-057-ME(2). Orono, ME: University of Maine - Land and Water Resources Center.

Lubin JH, Boice Jr JD. 1997. Lung cancer risk from residential radon: meta-analysis of eight epidemiologic studies. Journal of the National Cancer Institute 89(1):49-57.

Lubin JH, Boice Jr JD, Edling C, Hornung RW, Howe G, Kunz E, Kusiak RA, Morrison HI, Radford EP, Samet JM, Tirmarche M, Woodward A, Yao SX, Pierce DA. 1994. Lung Cancer and Radon: A Joint Analysis of 11 Underground Miners Studies. Publication No. 94-3644. Bethesda, MD: National Institutes of Health.

Luftig SD, Weinstock L. 1997. Establishment of Cleanup Levels for CERCLA Sites with Radioactive Contamination. Directive 9200.4-18. Washington, DC: Environmental Protection Agency, Office of Solid Waste and Emergency Response.

Marcinowski F, Lucas RM, Yeager WM. 1994. National and regional distributions of airborne radon concentrations in U.S. homes. Health Physics 66(6):699-706.

Martell EA. 1974. Radioactivity of tobacco trichromes and insoluble cigarette smoke particles. Nature 249(5454):215-217.

Martinez-Aguirre A, Garcia-Orellana L, Garcia-Leon M. 1996. Enhanced U and Th concentrations in soils from a wet marshland washed by contaminated riverwaters. Applied Radiation Isotopes 47(9,10):1081-1087.

Means JL, Crerar DA, Duguid JO. 1978. Migration of radioactive waste: radionuclide mobilization by complexing agents. Science 200(4349):1477-1481.

Menetrez MY, Watson JE. 1983. Natural Radioactivity in North Carolina Ground Water Supplies. WRRI-UNC-83-208. Chapel Hill, NC: University of North Carolina, Water Resources Research Institute.

Meriwether JR, Burns SF, Thompson RH, Beck JN. 1995. Evaluation of soil radioactivities using pedologically based sampling techniques. Health Physics 69(3):406-409.

Minerals Management Service. 1996a. Environmental Assessment for Issuance of Notice to Lessees and Operators of Federal Oil and Gas Leases on the Outer Continental Shelf, Gulf of Mexico Region: Guidelines for the Offshore Storage and Sub-Seabed Disposal of Waste Resulting from the Development and Production of Oil and Gas on the Outer Continental Shelf. Herndon, VA: Minerals Management Service.

Minerals Management Service. 1996b. Notice to Lessees and Operators of Federal Oil and Gas Leases on the Outer Continental Shelf, Gulf of Mexico Region: Guidelines for the Offshore Storage and Sub-Seabed Disposal of Waste Resulting from the Development and Production of Oil and Gas on the Outer Continental Shelf. NTL 96-03. Herndon, VA: Minerals Management Service.

Myrick TE, Berven BA, Haywood FF. 1981. State Background Radiation Levels: Results of Measurements Taken During 1975-1979. ORNL/TM-7343. Oak Ridge, TN: Oak Ridge National Laboratory.

National Research Council. 1983. Risk Assessment in the Federal Government: Managing the Process. Washington, DC: National Academy Press.

National Research Council. 1988. Health Risks of Radon and Other Internally Deposited Alpha-Emitters (BEIR IV). Washington, DC: National Academy Press.

National Research Council. 1990. Health Effects of Exposure to Low Levels of Ionizing Radiation (BEIR V). Washington, DC: National Academy Press.

National Research Council. 1991. Comparative Dosimetry of Radon in Mines and Homes. Washington, DC: National Academy Press.

National Research Council. 1995. Technical Bases for Yucca Mountain Standards. Washington, DC: National Academy Press.

Nazaroff WW, Doyle SM, Nero AV, Sextro RG. 1987. Potable water as a source of airborne Rn-222 in U.S. dwellings: a review and assessment. Health Physics 52(3):281-295.

Nazaroff WW, Nero AV. 1988. Radon and Its Decay Products in Indoor Air. New York: John Wiley and Sons.

NCRP. (National Council on Radiation Protection and Measurements). 1984a. Control of Air Emissions of Radionuclides. Bethesda, MD: National Council on Radiation Protection and Measurements.

NCRP. (National Council on Radiation Protection and Measurements). 1984b. Evaluation of Occupational and Environmental Exposures to Radon and Radon Daughters in the United States. NCRP Report No. 78. Bethesda, MD: National Council on Radiation Protection and Measurements.

NCRP. (National Council on Radiation Protection and Measurements). 1984c. Exposures from the Uranium Series with Emphasis on Radon and its Daughters. NCRP Report No. 77. Bethesda, MD: National Council on Radiation Protection and Measurements.

NCRP. (National Council on Radiation Protection and Measurements). 1984d. Radiological Assessment: Predicting the Transport, Bioaccumulation, and Uptake by Man of Radionuclides Released to the Environment. NCRP Report No. 76. Bethesda, MD: National Council on Radiation Protection and Measurements.

NCRP. (National Council on Radiation Protection and Measurements). 1987a. Exposure of the Population in the United States and Canada from Natural Background Radiation. NCRP Report No. 94. Bethesda, MD: National Council on Radiation Protection and Measurements.

NCRP. (National Council on Radiation Protection and Measurements). 1987b. Radiation Exposure of the U.S. Population from Consumer Products and Miscellaneous Sources. NCRP Report No. 95. Bethesda, MD: National Council on Radiation Protection and Measurements.

NCRP. (National Council on Radiation Protection and Measurements). 1987c. Recommendations on Limits for Exposure to Ionizing Radiaton. NCRP Report No. 91. Bethesda, MD: National Council on Radiation Protection and Measurements.

NCRP. (National Council on Radiation Protection and Measurements). 1988. Measurement of Radon and Radon Daughters in Air. NCRP Report No. 97. Bethesda, MD: National Council on Radiation Protection and Measurements.

NCRP. (National Council on Radiation Protection and Measurements). 1989a. Control of Radon in Houses. NCRP Report Report 103. Bethesda, MD: National Council on Radiation Protection and Measurements.

NCRP. (National Council on Radiation Protection and Measurements). 1989b. Guidance on Radiation Received in Space Activities. NCRP Report No. 98. Bethesda, MD: National Council on Radiation Protection and Measurements.

NCRP. (National Council on Radiation Protection and Measurements). 1993a. Limitation of Exposure to Ionizing Radiation. NCRP Report No. 116. Bethesda, MD: National Council on Radiation Protection and Measurements.

NCRP. (National Council on Radiation Protection and Measurements). 1993b. Radiation Protection in the Mineral Extraction Industry. NCRP Report No. 118. Bethesda, MD: National Council on Radiation Protection and Measurements.

NCRP. (National Council on Radiation Protection and Measurements). 1995. Radiation Exposure and High Altitude Flight. NCRP Commentary No. 12. Bethesda, MD: National Council on Radiation Protection and Measurements.

Neiheisel J. 1990. Characterization of soil contaminants for remedial measures. In: EPA Workshop on Radioactively Contaminated Sites. EPA 520/1-90-009. Washington, DC: Environmental Protection Agency. pp. 42-50.

Nero Jr AV. 1988. Radon and its decay products in indoor air: an overview. In: Nazaroff WW, Nero AV (eds.), Radon and Its Decay Products in Indoor Air. New York: John Wiley and Sons. pp. 1-56.

Nero Jr AV, Gadgil AJ, Nazaroff WW, Revzan KL. 1990. Indoor Radon and Decay Products: Concentrations, Causes and Control Strategies. DOE/ER-0480P. Washington, DC: Department of Energy.

Nero Jr AV, Schwehr MB, Nazaroff WW, Revzan KL. 1986. Distribution of airborne radon-222 concentrations in U.S. houses. Science 234(4779):992-997.

NIRH. (National Institute of Radiation Hygiene). 1987. Natural Radiation in Danish Dwellings. Copenhagen, Denmark: National Institute of Radiation Hygiene.

Nordic Radiation Protection Institutes. 1986. Naturally Occurring Radiation in the Nordic Countries Recommendations.

Nuclear Regulatory Commission. 1974. Termination of Operating Licenses for Nuclear Reactors. Regulatory Guide 1.86. Washington, DC: Nuclear Regulatory Commission.

Nuclear Regulatory Commission. 1977. Calculation of Annual Doses to Man from Routine Releases of Reactor Effluents for the Purpose of Evaluating Compliance with 10 CFR Part 50, Appendix I. Regulatory Guide 1.109, Rev. 1. Washington, DC: Nuclear Regulatory Commission.

Nuclear Regulatory Commission. 1981. Disposal or Onsite Storage of Residual Thorium or Uranium (Either as Natural Ores or Without Daughters Present) from Past Operations. Branch Technical Position. SECY-81-576. Washington, DC: Nuclear Regulatory Commission.

Nuclear Regulatory Commission. 1982a. 10 CFR Parts 2 et al.—Licensing requirements for land disposal of radioactive waste. Final rule. Federal Register 47(248):57446-57482.

Nuclear Regulatory Commission. 1982b. Guidelines for Decontamination of Facilities and Equipment Prior to Release for Unrestricted Use or Termination of Licenses for Byproduct, Source or Special Nuclear Material. Washington, DC: Nuclear Regulatory Commission.

Nuclear Regulatory Commission. 1991. 10 CFR Part 20—Standards for protection against radiation. Final rule. Federal Register 56(98):23360-23474.

Nuclear Regulatory Commission. 1997a. 10 CFR Part 20—Radiological criteria for license termination. Final rule. Federal Register 62(139):39058-39092.

Nuclear Regulatory Commission. 1997b. Draft Guidance on Radioactive Materials in Sewage Sludge/Ash at Publicly Owned Treatment Works (POTWS). Washington, DC: Nuclear Regulatory Commission.

Nuclear Regulatory Commission. 1997c. 10 CFR Parts 20 and 40—Radiological criteria for license termination: uranium recovery facilities. Proposed rule. Federal Register 62(139):39093-39095.

Nuclear Regulatory Commission/EPA. (Nuclear Regulatory Commission and Environmental Protection Agency). 1995. White Paper on Risk Harmonization. Washington, DC: Nuclear Regulatory Commission and Environmental Protection Agency (September).

Nuclear Regulatory Commission/EPA. (Nuclear Regulatory Commission and Environmental Protection Agency). 1996. Multi-Agency Radiation Survey and Site Investigation Manual (MARSSIM). NUREG-1575, EPA 402-R-96-018. Washington, DC: Nuclear Regulatory Commission.

O'Brien K, Friedberg W, Duke FE, Snyder L, Darden Jr EB, Sauer HH. 1992. The exposure of aircraft crews to radiations of extraterrestrial origin. Radiation Protection Dosimetry 45(1/4):145-162.

O'Brien K, McLaughlin JE. 1972. The radiation dose to man from galactic cosmic rays. Health Physics 22:225-232.

Oakley DT. 1972. Natural Radiation Exposure in the United States. ORD/SID 72-1. Washington, DC: Environmental Protection Agency.

Oge M. 1992. Overview of the United States Environmental Protection Agency radon action programme. Radiation Protection Dosimetry 45(1/4):751-757.

OSTP. (Office of Science and Technology Policy). 1985. Chemical carcinogens: a review of the science and its associated principles. Federal Register 50(50):10372-10442.

Overy DP, Richardson ACB. 1995. Regulation of radiological and chemical carcinogens: current steps toward risk harmonization. Environmental Law Reporter 25(12):10655-10708.

Paredes CH, Kessler WV, Landolt RR, Ziemer PL, Paustenbach DJ. 1987. Radionuclide content of and Rn-222 emanation from building materials made from phosphate industry waste products. Health Physics 53(1):23-29.

Paschoa AS. 1993. Overview of environmental and waste management aspects of the monazite cycle. Radiation Protection in Australia 11:170-173.

Paschoa AS. 1994. The monazite cycle in Brazil: past, present and future. In: Mishra B, Averill WA (eds.), Actinide Processing: Methods and Materials. Warrendale, PA: The Minerals, Metals & Materials Society. pp. 323-338.

Paschoa AS. 1997a. Naturally occurring radioactive materials (NORM) and petroleum origin. Applied Radiation Isotopes 48(10-12):1391-1396.

Paschoa AS. 1997b. NORM and the Brazilian non-nuclear industries. In: Proceedings of the International Symposium on Radiological Problems with Natural Radioactivity in the Non-Nuclear Industry, September 8-10, 1997. Arnhem, The Netherlands: KEMA. Section 3.5.

Paschoa AS, Nogueira CA, Lourenco MC, Malmutt C, Wrenn ME. 1992. Local Auger electron dosimetry of ^{40}K. Radiation Protection Dosimetry 45(1/4):677-679.

Pennders RMJ, Koster HW, Lembrechts JF. 1992. Characteristics of ^{210}Po and ^{210}Pb in effluents from phosphate-producing industries: a first orientation. Radiation Protection Dosimetry 45(1/4):737-740.

Perel'man AI. 1977. Geochemistry of Elements in the Supergene Zone. Jerusalem: Israel Program for Scientific Translations.

Pershagen G, Liang Z, Hrubec Z, Svensson C, Boice J. 1992. Residential radon exposure and lung cancer in Swedish women. Health Physics 63(2):179-186.

Persson BR. 1972. Radiolead (210-Pb), polonium (210-Po) and stable lead in the lichen, reindeer, and man. In: Adams JAS, Lowder WM, Gesell TF (eds.), The Natural Radiation Environment II. CONF-720805-P2. Washington, DC: Energy Research and Development Administration. pp. 347-367.

Peter Gray & Associates. 1997. The NORM Report. Summer/Fall Issue. Tulsa, OK: Peter Gray & Associates.

Pfister H, Philipp G, Pauly H. 1976. Population dose from natural radionuclides in phosphate fertilizers. Radiation and Environmental Biophysics 13(3):247-261.

Prichard HM, Gesell TF. 1983. Radon-222 in municipal water supplies in the Central United States. Health Physics 45(5):991-993.

Puskin JS, Nelson CB. 1995. Estimates of radiogenic cancer risk. Health Physics 69(1):93-101.

Puskin JS, Nelson NS, Nelson CB. 1992. Bone cancer risk estimates. Health Physics 63(5):579-580.

Reitz G, Schnuer K, Shaw K. 1993. Proceedings of Workshop on Radiation Exposure of Civil Aircrew. Radiation Protection Dosimetry 48(1).

Remick FJ. 1992. Regulatory risk coherence. In: American Nuclear Society Topical Meeting on Risk Management—Expanding Horizons. S-20-92. Washington, DC: Nuclear Regulatory Commission, Office of Public Affairs.

Reynolds BR. 1995. Who's going to regulate NORM? Northern Kentucky Law Review 22 (Rev. 5).

Rittiger CL, Yusko JG. 1996. An overview of NORM data collected in Pennsylvania specific to oil and gas production. In: Mossman KL, Thiemann KB (eds.), NORM/NARM: Regulation and Risk Assessment, Proceedings of the 29th Midyear Topical Meeting of the Health Physics Society. McLean, VA: Health Physics Society. pp. 129-132.

Schaefer HJ. 1971. Radiation exposure in air travel. Science 173(3999):780-783.

Schmitz J, Fritsche R. 1992. Radon impact at underground workplaces in western Germany. Radiation Protection Dosimetry 45(1/4):193-195.

Schoenberg JB, Lotz JB, Wilcox HB, Nicholls GP, Gil del Real MT, Stemhagen A, Mason TJ. 1990. Case-control study of residential radon and lung cancer among New Jersey women. Cancer Research 50(20):6520-6524.

Scott HL. 1992. Initial studies on levels of indoor radon at U.S. Department of Energy facilities. Health Physics 62(Suppl. 6):559.

Simon SL, Ibrahim SA. 1990. Biological uptake of radium by terrestrial plants. In: The Environmental Behaviour of Radium. Technical Report Series No. 310, Vol. 1. Vienna: International Atomic Energy Agency. pp. 545-599.

Simon V. 1990. Ottawa radiation sites. In: EPA Workshop on Radioactively Contaminated Sites, May 3-5 1989, Albuquerque, NM. EPA 520/1-90-009. Washington, DC: Environmental Protection Agency. pp. 66-68.

Smith KP. 1992. An Overview of Naturally Occurring Radioactive Materials (NORM) in the Petroleum Industry. ANL/EAIS-7. Argonne, IL: Argonne National Laboratory.

Spezzano E. 1993. Determination of alpha-emitting nuclides of uranium, thorium and radium in zircon sands. Radiation Protection in Australia 11:117-121.

Stannard JN. 1988. Radioactivity and Health: A History. DOE/RL/01830-T59. Richland, WA: Pacific Northwest Laboratory.

Steinhäusler FA. 1975. Long-term measurements of Rn-222, Rn-220, Pb-214 and Pb-212 concentrations in the air of private and public buildings and their dependence on meterological parameters. Health Physics 29:705-713.

Stohr JS, Erickson JL. 1984. Regulation of a post-glacial uranium deposit in the state of Washington. In: Proceedings of the Sixth Symposium on Uranium Mill Tailings Management. Ft. Collins, Co: Colorado State University, Geotechnical Engineering Program. pp. 15-23.

Straub CP. 1971. Radioactivity. In: American Water Works Association, Water Quality and Treatment—A Handbook of Public Water Supplies. New York: McGraw Hill. pp. 441-462.

Sullivan RE, Nelson NS, Ellett WH, Dunning Jr DE, Leggett RW, Yalcintas MG, Eckerman KF. 1981. Estimates of Health Risk from Exposure to Radioactive Pollutants. ORNL/TM-7745. Oak Ridge, TN: Oak Ridge National Laboratory and Environmental Protection Agency.

Summerlin Jr J, Prichard HM. 1985. Radiological health implications of lead-210 and polonium-210 accumulations in LPG refineries. American Industrial Hygiene Association Journal 46(4):202-205.

Tanner AB. 1964. Radon migration in the ground: a review. In: Adams JAS, Lowder WM (eds.), The Natural Radiation Environment. Chicago, IL: University of Chicago Press. pp. 161-190.

Tanner AB. 1980. Radon migration in the ground: a supplementary review. In: Gesell TF, Lowder WM (eds.), Proceedings of The Natural Radiation Environment III. CONF-780422. Springfield, VA: National Technical Information Service. pp. 5-56.

Tanner AB. 1992. Bibliography of Radon in the Outdoor Environment and Selected References on Gas Mobility in the Ground. Geological Survey Open-File 92-351. Denver, CO: US Geological Survey.

Taylor JM. 1995. U.S. Nuclear Regulatory Commission and U.S. Environmental Protection Agency Risk Harmonization Issues and Recommendations. SECY-95-249. Washington, DC: Nuclear Regulatory Commission.

Till JE, Meyer HR. 1983. Radiological Assessment: A Textbook on Environmental Dose Analysis. NUREG/CR-3332, ORNL-5986. Washington, DC: Nuclear Regulatory Commission.

Travis CC, Richter SA, Crouch EAC, Wilson R, Klema ED. 1987. Cancer risk management: a review of 132 federal regulatory decisions. Environmental Science & Technology 21(5):415-420.

Trovato R. 1997. Statement on the Nuclear Regulatory Commission's Rule on Radiological Critieria for License Termination. Washington, DC: Environmental Protection Agency, Office of Radiation and Indoor Air.

Turk BH, Brown JT, Geisling-Sobotka F, Grimsrud DT, Harrison J, Koonce JF, Revzan KL. 1986. Indoor Air Quality and Ventilation Measurements in 38 Pacific Northwest Commercial Buildings. LBL-21453. Berkeley, CA: Lawrence Berkeley Laboratory.

UNSCEAR. (United Nations Scientific Committee on the Effects of Atomic Radiation). 1982. Ionizing Radiation: Sources and Biological Effects. New York: United Nations.

UNSCEAR. (United Nations Scientific Committee on the Effects of Atomic Radiation). 1986. Ionizing Radiation: Sources and Biological Effects. New York: United Nations.

UNSCEAR. (United Nations Scientific Committee on the Effects of Atomic Radiation). 1988. Sources, Effects and Risks of Ionizing Radiation. New York: United Nations.

UNSCEAR. (United Nations Scientific Committee on the Effects of Atomic Radiation). 1993. Sources and Effects of Ionizing Radiation. New York: United Nations.

Upton AC, Chase HB, Hekhius GL. 1966. Radiobiological aspects of the supersonic transport: a report of the ICRP Task Group on the biological effects of high energy radiations. Health Physics 12:209-226.

van Brederode LE, Bosnjakovic BFM. 1992. The Dutch environmental policy on indoor radiation protection. Radiation Protection Dosimetry 45(1/4):771-774.

VandenBygaart AJ, Protz R. 1995. Gamma radioactivity on a chronosequence, Pinery Provincial Park, Ontario. Canadian Journal of Soil Science 75(1):73-84.

Venuti CG, Piermattei S. 1992. The impact of adopting EC recommendations on indoor radon. Radiation Protection Dosimetry 45(1/4):767-770.

Watson Jr JE, Mitsch BF. 1987. Groundwater concentrations for Ra-226 and Rn-222 in North Carolina phosphate lands. Health Physics 52(3):361-365.

Weimer WC, Kinnison RR, Reeves JH. 1981. Survey of Radionuclide Distributions Resulting from the Church Rock, New Mexico, Uranium Mill Tailings Pond Dam Failure. NUREG/CR-2449. Washington, DC: Nuclear Regulatory Commission.

White SB, Bergsten JW, Alexander BV, Rodman NF, Phillips JL. 1992. Indoor Rn-222 concentrations in a probability sample of 43,000 houses across 30 states. Health Physics 62(1):41-50.

Wilkening M. 1952. Natural radioactivity as a tracer in the sorting of aerosols according to mobility. Review of Scientific Instruments 23(1):13-16.

Wilkening MH, Watkins DE. 1976. Air exchange and Rn-222 concentrations in the Carlsbad Caverns. Health Physics 31:139-145.

Wolbarst AB, Mauro J, Anigstein R, Back D, Bartlett JW, Beres D, Chan D, Clark ME, Doehnert M, Durman E, Hay S, Hull HB, Lailas N, MacKinney J, Ralson L, Tsirigotis PL. 1996. Technical basis for EPA's proposed regulations on the cleanup of sites contaminated with radioactivity. Health Physics 71(5):644-660.

Wrenn ME, Singh NP, Cohen N, Ibrahim SA, Saccomano G. 1981. Thorium in Human Tissues. NUREG/CR-1227. Washington, DC: Nuclear Regulatory Commission.

Yarborough KA. 1980. Radon- and thoron-produced radiations in National Park Service caves. In: Gesell TF, Lowder WM (eds.), The Natural Radiation Environment III. Washington, DC: Department of Energy. pp. 1371-1395.

Yusko JG. 1997. An overview of Pennsylvania's experience with NORM. In: Barkenbus J, Gresalfi M (eds.), Beneficial Reuse '96, The Fourth Annual Conference on the Recycle and Reuse of Radioactive Scrap Metal. Knoxville, TN: University of Tennessee.

Appendix

Radiation Quantities and Units
Definitions
Acronyms

The quantities and units of radiation dose are inherently more complex than those used in toxicology or pharmacology, and additional complexity has resulted from several changes required by evolving concepts in radiation dosimetry. The first widely used physical quantity of radiation was "exposure," related to the ability of x or gamma radiation to ionize air; its unit was the roentgen (R). Exposure was limited to photon radiation with energy less than 2.5 MeV. The quantity "absorbed dose" (D) was introduced because it was applicable to all forms of ionizing radiation and absorbing materials. Absorbed dose is energy deposited per unit mass, and its original unit was the rep (roentgen equivalent-physical); 1 rep equaled 93 ergs per g (0.0093 J per kg) of absorbing material. The rep was replaced with the rad (radiation absorbed dose); 1 rad equaled 100 ergs per g (0.01 J per kg). The "dose equivalent" (H) and its unit, the rem (roentgen equivalent-man), were introduced to account for the different biologic effects of the same absorbed dose from different types of radiation; H is the product of D, Q, and N at a point of interest in tissue, where D is absorbed dose, Q is the quality factor, and N is the product of any other modifying factors. The "effective dose equivalent" was introduced to include the different sensitivities of individual tissues and organs, which are important for internal dosimetry: its unit is the same as the unit of "dose equivalent."

In the 1990 recommendations of the International Commission on Radiological Protection (ICRP 1991), the use of N was dropped and the radiation weighting factor (w_R) was substituted for Q. In addition, Systeme International (SI) units have been adopted by ICRP (1977). The unit of dose is

APPENDIX 271

now the gray (Gy), and the unit of equivalent dose, effective dose, and associated quantities is the sievert (Sv). Each of those units equals 1 J per kg. In terms of conventional units, 1 Gy = 100 rad and 1 Sv = 100 rem.

SI units have been almost universally adopted internationally and in the US scientific community, but they have not been embraced enthusiastically by the US regulatory and engineering communities. The principal international authority on radiologic quantities and units is the International Commission on Radiation Units and Measurements (ICRU), which maintains administrative offices in the National Council on Radiation Protection and Measurements (NCRP) headquarters in Bethesda, Maryland.

Definitions of various terms, quantities, and units used to describe radioactivity, radiation, and their control are given below. Most have been adapted from "Standards for Protection Against Radiation," Title 10, Part 20, of the Code of Federal Regulations (10 CFR 20). Definitions of effective dose and equivalent dose were adapted from ICRP (1991).

DEFINITIONS

absorbed dose The energy imparted by ionizing radiation per unit mass of irradiated material. The units of absorbed dose are the rad and the gray (Gy).

activity The rate of disintegration (transformation) or decay of radioactive material. The units of activity are the curie (Ci) and the becquerel (Bq).

As Low As Reasonably Achievable (ALARA) – Making every reasonable effort to maintain exposures as far below the dose limits as is practical, taking into account economic considerations and other societal concerns.

becquerel (Bq) The SI unit of activity. 1 Bq equals 1 disintegration per second.

byproduct material As used in the Atomic Energy Act:

(1) Any radioactive material (except special nuclear material) yielded in, or made radioactive by, exposure to the radiation incident to the process of producing or using special nuclear material; and
(2) The tailings or wastes produced by the extraction or concentration of uranium or thorium from ore processed primarily for its source-material content, including discrete surface wastes resulting from uranium-solution extraction processes. Underground ore bodies depleted by solution-

extraction operations do not constitute byproduct material according to this definition.

collective dose The sum of the individual doses received in a given period by a specified population from exposure to a specified source of radiation.

committed dose equivalent ($H_{T,50}$) The dose equivalent to organs or tissues of reference (T) that will be received from an intake of radioactive material by a person during the 50-y period after the intake.

committed effective dose equivalent ($H_{E,50}$) The sum of the products, for each body organ or tissue that is irradiated, of the applicable weighting factor and the committed dose equivalent to the organ or tissue ($H_{E,50} = \Sigma w_T H_{T,50}$).

curie (Ci) The conventional unit of activity. 1 Ci equals 3.7×10^{10} disintegrations per second, which equals 3.7×10^{10} Bq.

dose or **radiation dose** A generic term that means absorbed dose, dose equivalent, effective dose equivalent, committed dose equivalent, committed effective dose equivalent, or total effective dose equivalent.

dose equivalent (H_T) The product of the absorbed dose in tissue, the quality factor, and all other necessary modifying factors at the location of interest. The units of dose equivalent are the rem and the sievert (Sv).

effective dose (E) The sum of weighted equivalent doses to all tissues and organs of the body (ICRP 1991). $E = \Sigma w_T H_T$, where H_T is the equivalent dose and w_T is the tissue weighting factor.

effective dose equivalent (H_E) The sum of the products, for each body organ or tissue that is irradiated, of the dose equivalent to the organ or tissue and the applicable weighting factor ($H_E = \Sigma w_T H_T$).

equivalent dose (H_T) In radiation protection, the absorbed dose averaged over a tissue or organ (rather than a point, as is the case for dose equivalent) and weighted for the radiation quality that is of interest. For this quantity, the weighting factor is called the radiation weighting factor instead of the quality factor, as used in earlier dosimetric quantities.

external dose The dose received from radiation sources outside the body.

exposure A quantity used to express external ionizing radiation, or to indicate presence of radionuclides or radiation affecting individuals or populations (for example, "exposure" to radionuclides in the environment).

gray (Gy) The SI unit of absorbed dose. 1 Gy equals an absorbed dose of 1 J/kg (100 rad).

internal dose The dose received from radioactive material taken into the body.

limits (dose limits) The permissible upper bounds of radiation doses.

member of the public Any person except when that person is receiving an occupational dose.

quality factor (Q) The modifying factor that is used to derive dose equivalent from absorbed dose for purposes of radiation protection.

rad The special unit of absorbed dose. 1 rad equals an absorbed dose of 100 ergs per gram or 0.01 J per kg (0.01 Gy).

radiation (ionizing radiation) Alpha particles, beta particles, gamma rays, x rays, neutrons, high-energy electrons, high-energy protons, and other particles capable of producing ionization in matter. (As used in this report, radiation does not include nonionizing radiation, such as radiowaves, microwaves, visible, infrared, or ultraviolet light.)

reference man A hypothetical aggregation of human physical and physiologic characteristics arrived at by international consensus. These characteristics can be used by researchers and public-health workers to standardize results of experiments and to relate biologic insult to a common base.

relative biological effectiveness (RBE) The ratio of the absorbed dose of a reference radiation (usually 200 keV x rays) to the absorbed dose of the test radiation required to produce the same degree of biologic effect. The RBE of the test radiation depends on the exact biologic effect in a given species of organism under a given set of exposure conditions.

rem The special unit of any of the quantities expressed as dose equivalent. The dose equivalent equals the product of the absorbed dose in rads and the quality factor (1 rem = 0.01 Sv).

roentgen (R) The unit of exposure. One roentgen equals the amount of x or gamma radiation required to produce ions carrying a charge of 1 electrostatic unit (esu) per cubic centimeter (2.58 x 10^{-4} coulomb per kg) of dry air under standard conditions.

sievert (Sv) The SI unit of any of the quantities expressed as dose equivalent. The dose equivalent in sieverts is equal to the product of the absorbed dose in grays and the quality factor (1 Sv = 100 rem).

source material As defined under the Atomic Energy Act:

(1) Uranium, thorium, or any combination of uranium and thorium in any physical or chemical form; or
(2) Ores that contain, by weight, 0.05% or more of uranium, thorium, or any combination thereof. Source material does not include special nuclear material.

special nuclear material As defined under the Atomic Energy Act:

(1) Plutonium, uranium-233, uranium enriched in uranium-233 or in uranium-235, and any other material that the Nuclear Regulatory Commission, pursuant to the provisions of Section 51 of the act, determines to be special nuclear material, but not including source material; or
(2) Any material artificially enriched by any of the foregoing, but not including source material.

Total Effective Dose Equivalent (TEDE) The sum of the deep-dose equivalent (for external exposures) and the committed effective dose equivalent (for internal exposures). It is a term used by some organizations to emphasize that the sum of the contributions from external and internal sources is meant. This term is not a part of the recommendations of the ICRP or NCRP. The term *effective dose equivalent*, without the modifier "total," is sufficient to imply contributions from external and internal sources.

uranium fuel cycle The operations of milling of uranium ore, chemical conversion of uranium, isotopic enrichment of uranium, fabrication of uranium fuel, generation of electricity by a light-water-cooled nuclear power plant using uranium fuel, and reprocessing of spent uranium fuel to the extent that these activities directly support the production of electric power for public use. Does not include mining operations, operations at waste-disposal sites, transportation of radioactive material in support of these operations, and the reuse of recovered nonuranium special nuclear and byproduct materials from the cycle.

weighting factor (w_T) For an organ or tissue (T), the proportion of the risk of stochastic effects resulting from irradiation of that organ or tissue to the total risk of stochastic effects when the whole body is irradiated uniformly.

whole body For purposes of external exposure, the head, trunk (including male gonads), arms above the elbow, and legs above the knee.

working level (WL) Any combination of short-lived radon decay products in 1 L of air that will result in the ultimate emission of alpha-particle energy equal to 1.3×10^5 MeV (2.08×10^{-5} J per m^3). Also equals the total energy emitted by alpha particles from short-lived radon decay products in equilibrium with radon gas in air at a concentration of 100 pCi/L (3.7 kBq per m^3).

LIST OF ACRONYMS

AEC	Atomic Energy Commission
ALARA	As low as reasonably achievable
AK	Alaska
ARAR	Applicable or relevant and appropriate requirement
BEIR	Biological Effects of Ionizing Radiations
BSEE	Bachelor of Science in Electrical Engineering
CEC	Commission of the European Communities
CERCLA	Comprehensive Environmental Response, Compensation, and Liability Act
CFR	Code of Federal Regulations
CRCPD	Conference of Radiation Control Program Directors
DHHS	Department of Health and Human Services
DOE	Department of Energy
EC	European Communities
EDTA	Ethylenediamine tetraacetic acid
EPA	Environmental Protection Agency
ERR	Excess relative risk
FR	Federal Register
FRC	Federal Radiation Council
FUSRAP	Formerly Utilized Sites Remedial Action Program
GAO	Government Accounting Office
HI	Hawaii
HPS	Health Physics Society
IAEA	International Atomic Energy Agency

ICRP	International Commission on Radiological Protection
ICRU	International Commission on Radiation Units and Measurements
MCL	Maximum contaminant level
MCLG	Maximum contaminant level goal
MTHM	Metric tons of heavy metal
NARM	Naturally occurring and accelerator-produced radioactive materials
NCP	National contingency plan
NCRP	National Council on Radiation Protection and Measurements
NESHAP	National Emission Standards for Hazardous Air Pollutants
NJ	New Jersey
NORM	Naturally occurring radioactive materials
NPDES	National pollutant discharge elimination system
NRC	Nuclear Regulatory Commission
NYU	New York University
OSTP	Office of Science and Technology Policy
PG	Phosphogypsum
PIC	Pressurized ionization chamber
RCRA	Resource Conservation and Recovery Act
Rn	Radon
ROD	Record of Decision
RR	Relative risk
SAB	Science Advisory Board of the Environmental Protection Agency
SFMP	Surplus Facilities Management Program
SI	Systeme International (units)
SSRCR	Suggested state regulations for the control of radiation
TBC	To be considered
TEDE	Total effective dose equivalent
TENORM	Technologically enhanced naturally occurring radioactive materials
TLD	Thermoluminescent dosimeter
TSCA	Toxic Substances Control Act
UIC	Underground injection control
UK	United Kingdom
UNSCEAR	United Nations Scientific Committee on the Effects of Atomic Radiation
UMT	Uranium mill tailings
UMTRCA	Uranium Mill Tailings Radiation Control Act
US	United States
USDW	Underground source of drinking water

APPENDIX

US EPA	Environmental Protection Agency
WL	Working level
WLM	Working level month

Information on Committee Members

BERNARD D. GOLDSTEIN, M.D. (Chairman), is the Director of the Environmental and Occupational Health Sciences Institute, a joint program of Rutgers, the State University of New Jersey, and the University of Medicine and Dentistry of New Jersey (UMDNJ)-Robert Wood Johnson Medical School and the Chair of the Department of Environmental and Community Medicine, UMDNJ-Robert Wood Johnson Medical School. He is a physician, board certified in Internal Medicine and Hematology; board certified in Toxicology. He was Assistant Administrator for Research and Development, U.S. Environmental Protection Agency, 1983-1985. His past activities include Member and Chairman of the NIH Toxicology Study Section and EPA's Clean Air Scientific Advisory Committee; Chair of the Institute of Medicine Committee on the Role of the Physician in Occupational and Occupational/Environmental Medicine, the National Research Committee on Biomarkers in Environmental Health Research and the Committee on Risk Assessment Methodology. Dr. Goldstein also has served on the Industry Panel of the World Health Organization Commission on Health and Environment. He is a Member of the Institute of Medicine. He is Principal Investigator of Consortium for Risk Evaluation with Stakeholder Participation. He is the author of over two hundred articles and book chapters related to environmental health sciences and public policy.

MERRIL EISENBUD, B.S.E.E., D.Sc. (deceased, see *Dedication*) was known worldwide in the field of environmental radioactivity. He served 12 years (1947-1959) with the US Atomic Energy Commission and was the founding director of the Health and Safety Laboratory professor. Professor Eisenbud was director of the Laboratory of Environmental Studies at the New York University Medical Center's Institute of Environmental Medicine from 1959 until 1984. On retirement from active teaching at NYU in 1984 he continued as professor emeritus of environmental medicine. He was also distinguished scholar in residence at the Duke University Medical Center and adjunct professor of environmental sciences and engineering at the University of North Carolina School of Public Health. Professor Eisenbud held a BSEE from the New York University College of Engineering and two honorary doctoral degrees in science. He was a member of many national and international committees, including those of agencies of the United Nations, the National Research Council, and the US government. He had been a member of the advisory councils of the Electric Power Research Institute, the Institute of Nuclear Power Operations, and the Beryllium Industry Scientific Advisory Committee. He was serving the National Research Council as a member of its Board on Radiation Effects Research at the time of his death. He was the recipient of many awards including the Hermann M. Biggs Medal of the New York State Public Health Association, the Arthur H. Compton Award of the American Nuclear Society, the Gold Medal of the US Atomic Energy Commission, the Distinguished Achievement Award of the Health Physics Society, the Life Award of the Power Division of the Institute of Electronic and Electrical Engineers, and the Taylor Medal of the National Council on Radiation Protection and Measurements. He was an honorary life fellow of the New York Academy of Sciences, a member of the National Academy of Engineering, a corresponding member of the Brazilian Academy of Sciences, and a fellow of the New York Academy of Medicine. He contributed more than 200 journal articles and book chapters to the scientific literature.

THOMAS GESELL, Ph.D., is a Professor of Health Physics and the Director of the Technical Safety Office at Idaho State University. Dr. Gesell has worked in multiple capacities for the DOE Idaho Operations Office, including holding the position of Deputy Assistant Manager for Nuclear Programs, and was a faculty member of the University of Texas School of Public Health in Houston for ten years. He was the Director of the DOE Radiological and Environmental Sciences Laboratory on the Idaho National Engineering Laboratory Site. Dr. Gesell is a member of several committees and professional organizations including the EPA's Science Advisory Board's Radiation Advisory Committee and the National Council on Radiation Protection and Measurements. Dr. Gesell was also a consultant to the President's Commission on the Accident at Three Mile Island. Recently, Dr. Gesell co-authored the book *Environmental Radioactivity from Natural, Industrial and Military Sources* with Merril Eisenbud.

SHAWKI IBRAHIM, Ph.D., is an Associate Professor in the Department of Radiological Health Sciences at the Colorado State University. He received his Ph.D. degree in Environmental Health from New York University in 1980.

Formerly, he was a Research Scientist at New York University Medical Center's Laboratory for Environmental Medicine. Over the past 25 years, Dr. Ibrahim has been involved with research on measurements, distribution and transport of natural and man-made radionuclides in the environment and in humans around various nuclear facilities. He is a member of several professional organizations including the Health Physics Society, Sigma Xi, and the American Nuclear Society.

EDWARD LANDA, Ph.D., is a geochemist with the National Research Program of the USGS Water Resources Division. He holds a M.P.H. in radiological health, and an M.S. and Ph.D. in soil sciences from the University of Minnesota. His research has focused on radionuclide mobility in soil and aquatic environments, and has included studies of uranium mill tailings, radium processing residues, oil field brines, and indoor radon. Dr. Landa has served as the Department of the Interior representative to the Science Panel of the Committee on the Interagency Radiation Research and Policy Coordination from 1990 to 1995. He has participated in the IAEA International Chernobyl Project, and in studies of radionuclide contaminants in the Artic regions.

DAVID KOCHER, Ph.D., is a Senior Research Staff Member in the Life Sciences Division at Oak Ridge National Laboratory (ORNL). He earned his Ph.D. degree in physics from the University of Wisconsin. For the past two decades, he has worked as an environmental health physicist at ORNL. His principal research activities have involved the development of models and data bases for estimating radiation doses to the public from radionuclides in the environment, which have been widely used in assessing impacts of releases from operating facilities and from radioactive waste disposal, and evaluations of dose-assessment models for regulatory and decision-making purposes. He has served as a member of several technical advisory groups for the Department of Energy, the Science Advisory Board of the Environmental Protection Agency, the Nuclear Regulatory Commission, and the International Atomic Energy Agency in the areas of environmental radiological assessment and radioactive waste management. He is presently serving on Scientific Committees of the National Council on Radiation Protection and Measurements on risk-based waste classification and performance assessment for low-level waste disposal. He has lectured widely in the areas of external and internal dosimetry, radiological assessments, radiological assessments, radioactive waste management, and regulations for radionuclides and hazardous chemicals in the environment.

ANSELMO PASCHOA, Ph.D., is a Professor in the Department of Physics at the Pontifical Catholic University of Rio de Janeiro. Dr. Paschoa earned his Ph.D. from New York University. He has a broad background in physics including specialized training in nuclear and reactor physics, radiation dosimetry, and radioecology. Dr. Paschoa has been visiting professor at the University of Utah, and guest scientist at Brookhaven National Laboratory. Dr. Paschoa has been called upon by the Brazilian government to act as a consultant or committee member and has attended several international meetings as a representative of Brazil. He is former Director for Radiation Protection, Nuclear Safety, and Safeguards of the Brazilian Nuclear Energy Commission. Dr. Paschoa is also involved in numerous professional societies and organizations, and serves on the scientific committee for the Symposia on Radiation Physics.